Squeaky Clean Topology in Blender

Create accurate deformations and optimized geometry for characters and hard surface models

Michael Steppig

BIRMINGHAM—MUMBAI

Squeaky Clean Topology in Blender

Copyright © 2022 Packt Publishing

Group Product Manager: Rohit Rajkumar

Publishing Product Manager: Kaustubh Manglurkar

Senior Editor: Divya Anne Selvaraj

Technical Editor: Simran Haresh Udasi

Copy Editor: Safi s Editing

Project Coordinator: Sonam Pandey

Proofreader: Safi s Editing

Indexer: Subalakshmi Govindhan

Production Designer: Alishon Mendonca

Marketing Coordinator: Nivedita Pandey

First published: December 2022

Production reference: 1141222

Published by Packt Publishing Ltd.

Livery Place

35 Livery Street

Birmingham

B3 2PB, UK.

978-1-80324-408-2

www.packt.com

To my grade school English teacher, Joanne Heffernan, whose patience and love compelled me to write more than I ever wanted to.

– Michael Steppig

Contributors

About the author

Michael Steppig is a professional Blender user. He specializes in topology and character modeling. He started modeling in 2013 and has watched 3D modeling and Blender develop over the last decade. He has a strong passion for topology and ample experience, having modeled hundreds of 3D assets. These include characters, hard surface models, and nearly every asset he uses. He has taught multiple students all aspects of Blender as a private tutor.

I want to thank all of the people who helped me through the writing of this book – my family and friends, who all supported me when the writing was most difficult. I would also like to thank the team members at Packt who worked with me, as they were more than patient as I balanced this book with university. I could not have completed this book without their constant gentle support.

About the reviewer

Jesse Tomi is a self-taught Nigerian 3D generalist based in Abuja, Nigeria. He co-manages a start-up focused on creating seamless virtual experiences. Jesse is a graduate of educational technology from the University of Ilorin and has been involved with 3D graphics and animation for 5 years.

Table of Contents

6

Topology on a Humanoid Body   141

7

Topology on a Hard Surface   163

8

Optimizing Geometry for a Reduced Triangle Count   191

Preface

This book is an introduction to modeling and an in-depth look at topology in Blender, written by a Blender topology specialist with years of experience with the software. As you progress through its chapters, you'll conquer the basics of quad-based topology using triangles and Ngons, and learn best practices and things to avoid while modeling and retopologizing. The pages are full of illustrations and examples with in-depth explanations that showcase each step in an easy-to-follow format.

 Squeaky Clean Topology in Blender starts by introducing you to the user interface and navigation. It then goes through an overview of the modeling techniques and hotkeys that will be necessary to understand the examples. With the modeling basics out of the way, the next stop on our journey is topology. Working through projects like a character and a sci-fi blaster, the book will illustrate and work through complex topology problems, and present solutions to those problems. These examples focus on deforming character models, non-deforming hard surface models, and optimizing these models by reducing the triangle count.

By the end of this book, you will be able to identify the general flow of a shape's topology, identify and solve issues in your topology, and come out with a model ready for UV unwrapping, materials, and rigging.

Who this book is for

This book is for character modelers, sculptors, poly modelers, and hard surface modelers. Whether you're looking for an introduction to modeling and optimizing high-poly or sculpted models, or just a deeper dive into the subject of topology, this book will walk you through the topology workflow from beginning to end.

What this book covers

Chapter 1, *Navigating and Modeling in Blender*, will lay out the basic operation of modeling in Blender. It will start with navigation by providing a brief introduction to the UI, and then move on to explaining hotkeys and the basic modeling techniques used in the examples. Some hotkeys that I use specifically for topology will be discussed.

Chapter 2, *The Fundamentals of Topology*, will introduce you to a grid. We will understand how edges interact across a face and how they can be manipulated to create different shapes. I will provide examples of topology over different shapes and explain how separate faces interact.

Chapter 3, Deforming Topology, will explain how to place your topology to optimize deformations. The main deformations will be for soft body simulations and armatures. I will explain the densities, orientations, and overall shapes of specific joints.

Chapter 4, Improving Topology Using UV Maps, will explain how to lay out your topology to make unwrapping a mesh easier. First, we will explain how to position the topology. After that, we will show how to lay out the seams for the UV and how to actually unwrap them. Finally, we will look at the UV editing tab to identify potential issues with the topology or the seams.

Chapter 5, Topology on a Humanoid Head, will start with the organic retopology of a character's head. It will walk you through specific steps of the retopology process, giving you insight into what you should think about at every point. We will start by topologizing the eyes, then the nose, and finally, the mouth.

Chapter 6, Topology on a Humanoid Body, will start by identifying areas of detail and isolating them. These will act in the same way that the segments in the previous chapter did. They will be the anchors for us to build around – in this case, starting with the hands, then the arms, the shoulders, and finally, the hips.

Chapter 7, Topology on a Hard Surface, will focus on meshes that are not designed to deform. Because we are not deforming this mesh at all, we are only worried about shading artifacts. These are caused by the normals of the faces or the geometry itself getting messed up.

Chapter 8, Optimizing Geometry for a Reduced Triangle Count, is all about optimizing the triangle count of a mesh. This utilizes most of the tools and ideas we have been introduced to in the previous chapters. It is where we remove triangles wherever they are not needed. We do this by introducing triangles into deformable meshes and bending our previous topology rules.

To get the most out of this book

The latest version of Blender is the only technical requirement. A mouse is strongly advised, with many of the examples in this book referencing a mouse directly. A keyboard with a number pad is also advised but is less necessary.

Software/hardware covered in the book	Operating system requirements
Blender 3.2 or higher	Windows, macOS, or Linux

The latest version of Blender can be downloaded from `https://www.blender.org/download/`.

Download the color images

We also provide a PDF file that has color images of the screenshots and diagrams used in this book. You can download it here: `https://packt.link/5N7hA`.

Conventions used

There are some text conventions used throughout this book.

Bold: Indicates a new term, an important word, or words that you see on screen. For instance, words in menus or dialog boxes appear in **bold**. Here is an example: "Select **System info** from the **Administration** panel."

> **Tips or important notes**
> Appear like this.

Get in touch

Feedback from our readers is always welcome.

General feedback: If you have questions about any aspect of this book, email us at customercare@packtpub.com and mention the book title in the subject of your message.

Errata: Although we have taken every care to ensure the accuracy of our content, mistakes do happen. If you have found a mistake in this book, we would be grateful if you would report this to us. Please visit www.packtpub.com/support/errata and fill in the form.

Piracy: If you come across any illegal copies of our works in any form on the internet, we would be grateful if you would provide us with the location address or website name. Please contact us at copyright@packt.com with a link to the material.

If you are interested in becoming an author: If there is a topic that you have expertise in and you are interested in either writing or contributing to a book, please visit authors.packtpub.com.

Share Your Thoughts

Once you've read *Squeaky Clean Topology in Blender*, we'd love to hear your thoughts! Scan the QR code below to go straight to the Amazon review page for this book and share your feedback.

https://packt.link/r/1-803-24408-9

Your review is important to us and the tech community and will help us make sure we're delivering excellent quality content.

Download a free PDF copy of this book

Thanks for purchasing this book!

Do you like to read on the go but are unable to carry your print books everywhere?

Is your eBook purchase not compatible with the device of your choice?

Don't worry, now with every Packt book you get a DRM-free PDF version of that book at no cost.

Read anywhere, any place, on any device. Search, copy, and paste code from your favorite technical books directly into your application.

The perks don't stop there, you can get exclusive access to discounts, newsletters, and great free content in your inbox daily

Follow these simple steps to get the benefits:

1. Scan the QR code or visit the link below

https://packt.link/free-ebook/9781803238975

2. Submit your proof of purchase
3. That's it! We'll send your free PDF and other benefits to your email directly

Part 1 –
Getting Started with
Modeling and Topology

Topology is the base of all modeling operations. The quality of a model's topology has a direct effect on the quality and ease of use of the final model. Decisions regarding topology made at the early stages of a model can plague the whole modeling process and beyond, leaving problems that you may have to deal with every time you use the model. That is what makes an understanding of topology so imperative from the start.

We will be looking at topology in Blender, so a quick look at modeling will be our first objective. Once we have an understanding of general modeling in Blender, we will go straight into the fundamentals of topology. In each chapter, we will look at new rules to help guide us in the topology process. These are basic universal rules that apply to any software, with examples done in Blender.

The topics will include things to consider when modeling and then when deforming your model. Finally, we will cover what to think about when preparing your model for materials.

By the end of these chapters, you will have a general understanding of topology and what rules to follow when approaching the topology of a model. You will have learned checks that you can apply to models to determine the quality of their topology.

This part of the book comprises the following chapters:

1

Navigating and Modeling in Blender

Blender is a powerful 3D modeling software. It has a massive community and is frequently updated with new features that greatly improve the software. Since its support of photo-based materials and its UI update in *version 2.8*, Blender has become a legitimate professional option for 3D artists. It can perform most of the operations in the 3D modeling pipeline, such as sculpting, modelling, UV unwrapping, rigging, and much more. Blender is a massive generalist software that goes beyond just modeling and rigging. It also provides multiple physics engines, a compositor for video editing, and even 2D animation tools. However, the most impressive aspect Blender has going for it is not its impressive capabilities but, rather, the fact that it gives us these amazing capabilities and still remains completely free.

In this chapter, we will take a look at Blender's **user interface** (**UI**). We will learn how to navigate the different tabs that we will be using and the 3D viewport. Then, we will learn basic modeling techniques and hotkeys. Finally, we will take a look at specific tools used to check topology.

In this chapter, we will be learning the following subjects:

- Making sense of the Blender UI
- Navigating using the viewport
- What a 3D mesh is and how it can be manipulated
- Adding more vertices to enhance a mesh's geometry
- Manipulating a mesh using modifiers

Technical requirements

In this book, we will be using a vanilla *Blender* release, *version 3.2*, with included add-ons. All examples shown will be compatible with *version 3.3 LTS*, which is the latest version of Blender released at the time of writing. The latest version of Blender can be downloaded for free from `https://www.blender.org/download/`.

An external mouse will also be required, and a keyboard with a number pad is strongly recommended.

Making sense of the Blender UI

If this is your first time booting up Blender, it can be overwhelming – so many tabs and gizmos feeding you information. Of course, because you can do so many things in Blender, that also means there are a lot of options. So, before diving straight into topology, let's take a moment to look through the UI.

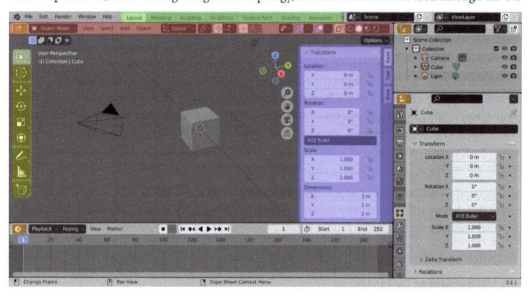

Figure 1.1 – The Blender UI

The Blender UI houses all of the different sections we will be using. The UI is broken up into five sections, the *Header*, *Workspaces*, *Areas*, the *Toolbar*, and the *Sidebar*. You can see these sections highlighted in *Figure 1.1*. Let us get a sense of what each of these areas contains:

- The **Header**, highlighted in red, contains **object interaction mode**, which determines how you want to affect an object, and the options for that mode. In the preceding figure, it is set to **Object Mode**.

- The **Workspaces**, highlighted in green, are preset tabs set up for a specific operation. In the preceding screenshot, we are in the **Layout** workspace.

- The **Areas** are where each of the individual **editor types**, highlighted in orange, are contained. Editor types are used to manage the many workflows of Blender.

- The **Toolbar**, highlighted in yellow, displays the tools for the object interaction mode that you have selected. Its visibility can be toggled by pressing *T*.

- The **Sidebar**, highlighted in blue, shows specific details about the object you have selected, such as its dimensions. It can be toggled by pressing *N*.

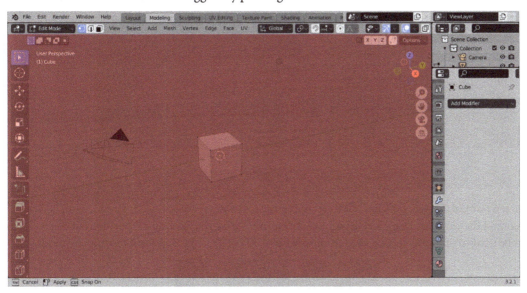

Figure 1.2 – The main region workspace

The workspace we will be working in the most will be the **Modeling** tab. The **main region**, highlighted in red in *Figure 1.2*, is the area that takes up the most screen space. The **Modelling** tab's main region is the 3D Viewport by default. The viewport is the editor type that shows the 3D object. This is the area that allows you to move around in 3D space. These are highlighted in *Figure 1.2*.

Now that we know where the major parts of the Blender UI are, we can take a look at how to navigate in the 3D Viewport. This is the main area we will be using when looking at our models.

Navigating using the viewport

While modeling, we spend most of our time in the viewport, so getting comfortable with its navigation is important. You navigate the viewport using three main avenues of movement, orbiting, zooming, and panning. To orbit, simply press the **middle mouse button** (**MMB**), and you will rotate around a hidden pivot point. To zoom, use the scroll wheel to scroll forward to zoom in and scroll backward to zoom out. To pan, press *Shift* + the *MMB* and move the mouse in the direction you want to pan.

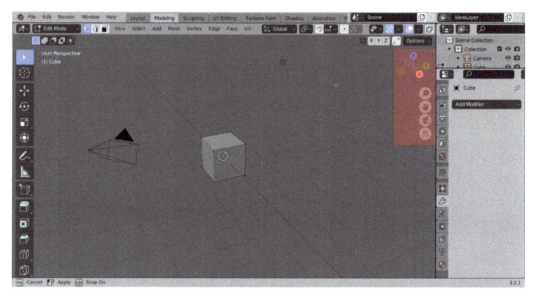

Figure 1.3 – Navigation gizmos

The top right of the viewport has a few helpful gizmos, as shown in *Figure 1.3*. These can all be used to control the viewport using just the mouse. The gizmo listing the **X**, **Y**, and **Z** positions in *Figure 1.4* can be used to orbit by pressing the **left mouse button** (**LMB**) on the icon and moving the mouse.

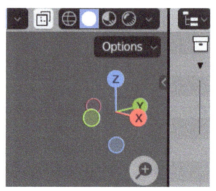

Figure 1.4 – The orbit gizmo

The magnifying glass icon below the orbiting gizmo in *Figure 1.5* can be used to zoom by pressing the *LMB* on the icon and moving the mouse.

Figure 1.5 – The zoom gizmo

And finally, the hand icon in *Figure 1.6* can be used to pan by pressing the *LMB* on the icon and moving the mouse.

Figure 1.6 – The pan gizmo

Now that we have some understanding of the viewport's gizmos and the features of Blender's UI, we can take a look at what we will be using these tools for.

What a 3D mesh is and how it can be manipulated

First, we will talk about what a 3D mesh actually is. A **mesh** constitutes one or more geometric points called vertices. These vertices can be connected by edges, and those edges can be connected with faces. A mesh is contained within an object. You can see all of these in *Figure 1.7*. An **object** contains one or more meshes and acts as a container for all of the data stored in the mesh.

Figure 1.7 – Vertices, edges, and faces forming a triangle

To edit the mesh of an object directly, you need to have the object you want to edit selected in the object selection mode called **Object Mode** (see *Figure 1.8*). To select an object, follow these steps:

1. Press the *LMB* on the object in the viewport.

2. Then, switch the object selection mode to **Edit Mode** by pressing *LMB* on the icon at the top left and selecting it from the drop-down list.

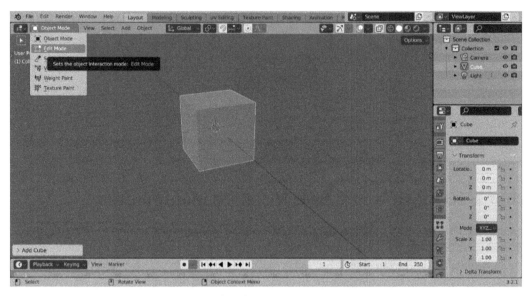

Figure 1.8 – The Object Interaction Mode tab

Alternatively, you can also press the *Tab* hotkey to switch to **Edit Mode** with the object selected.

Now that we are in **Edit Mode**, we can start to manipulate the mesh, starting with selection. *Select modes* are located in the top left of the header, as shown in *Figure 1.9*.

Figure 1.9 – Select modes – vertex, edge, and face from left to right

Select modes allow you to select a part of the mesh to change. The three buttons from left to right enable you to select a vertex, select an edge, and select a face respectively. You can enable these individually, or hold *Shift* and select multiple modes to enable.

To select a vertex, hover over the vertex you want to select and press the *LMB*. To select multiple vertices, hold *Shift* while selecting the other vertices. Selections have two types: selected and active. Selected vertices are a darker orange, while active vertices are highlighted in a brighter orange or white. *Figure 1.9* shows two vertices selected, with one of them active.

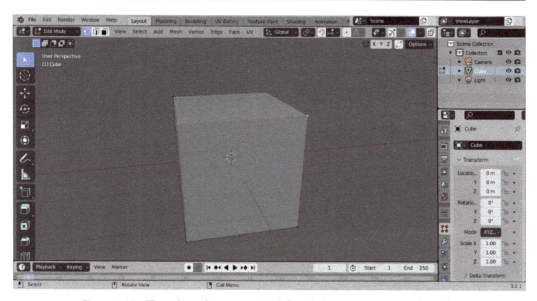

Figure 1.10 – The selected vertex on the left and the active vertex on the right

The active selection will always be the one that has its data displayed in the editor types and sidebar. You can press *A* to select everything and *Alt + A* to unselect everything.

Now that we know how to select different parts of the mesh, we can actually start doing operations

1. To translate a mesh, just select the vertices, edges, or faces you want, and then press *G* and move your mouse to the desired location.

2. It is equally easy to scale. Start by pressing *S*, and then move closer or farther from the point you are scaling from to change the scale.

3. To rotate, press *R* and rotate your cursor around the point or rotation to change the rotation.

4. To finalize any transform, simply press the *LMB*, and to cancel a transform, press the **right mouse button** (**RMB**).

5. You can also lock all of these transforms along a specific axis. Just press *X*, *Y*, or *Z* after pressing the transform button and it will lock you into its respective axis.

In *Figure 1.11*, you can see the mesh being scaled along just the *Z* axis by pressing *S* and then *Z*.

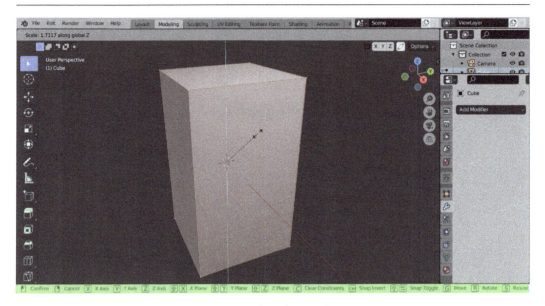

Figure 1.11 – Mesh scaling on the Z axis

Also, note that at the bottom of the screen, more transform options are listed, as highlighted in green in *Figure 1.11*. We will not be exploring these options within this book, but these can be helpful to know about when you start modeling on your own.

Now that we know how to move the mesh around, we can start adding some more geometry.

Adding more vertices to enhance a mesh's geometry

There are a few ways to add more vertices to a mesh. The one used most frequently later on in this book is extrusion. You can extrude a selection of a mesh by pressing *E* and moving the mouse to the desired extrusion point.

Figure 1.12 shows one vertex being extruded, although you can extrude faces, edges, and a bigger selection as well.

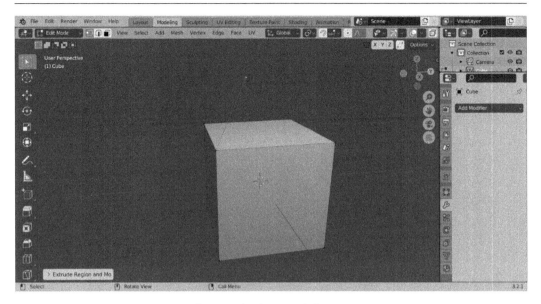

Figure 1.12 – An extruded vertex

Another way to create more geometry is to duplicate a mesh. To duplicate a mesh, just select the parts you want to duplicate and press *Shift + D*. This is usually the method used to start a new section of geometry. Just like the transforms listed earlier, to finalize the action, press the *LMB*, and to cancel it, press the *RMB*. Be careful when doing these operations, as right-clicking will just reset the location of the operation and will not delete the new geometry. *Figure 1.13* shows a face being duplicated.

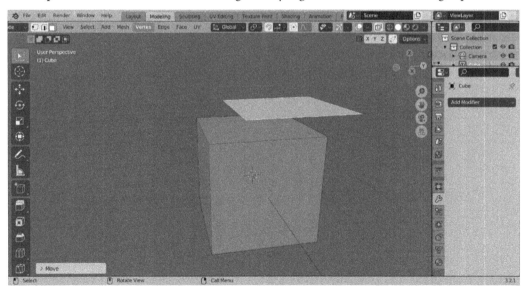

Figure 1.13 – A duplicated face

Once you have created a string of edges, as shown in *Figure 1.14*, you might want to create a face connecting them.

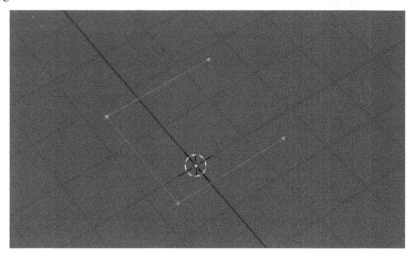

Figure 1.14 – Edges to make a face

To do this, you either need to select at least three vertices or two edges and press *F*. *Figure 1.15* shows the face that was created.

Figure 1.15 – A face made from the selected edges

We can also add geometry by subdividing the mesh. Subdividing the mesh divides the faces of a model into more small faces. To subdivide, select the faces that you want to subdivide and press the *RMB*. This will bring up the **Vertex Context Menu** options, as shown in *Figure 1.16*, with **Subdivide** at the top of the menu.

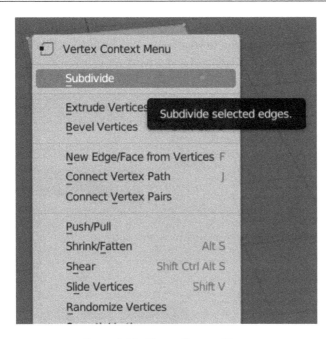

Figure 1.16 – Vertex Context Menu

After selecting **Subdivide**, your mesh will subdivide, and a new tab to control the subdivision will pop up in the bottom left. Both the mesh and the tab can be seen in *Figure 1.17*.

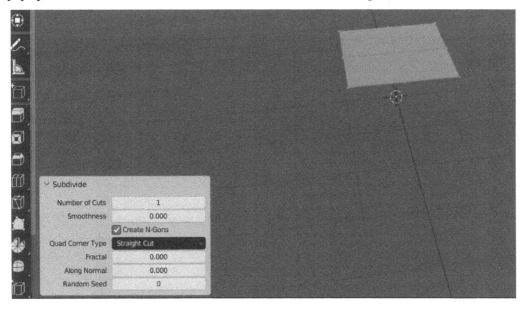

Figure 1.17 – The subdivided mesh and subdivision settings

You can also add any number of primitive objects to your scene. A **primitive object** is a basic shape such as a cube or sphere that you can add to your scene as a starting point for your model. To add a primitive in object mode, simply press *Shift + A* and hover your mouse over the mesh tab at the top, as shown in *Figure 1.18*.

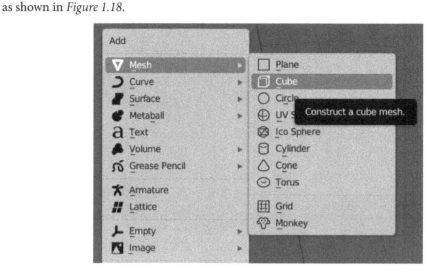

Figure 1.18 – The Add menu

Now that we know how to add geometry, we need to learn how to remove it as well. After selecting the desired areas for removal, press *X*. This will pull up a dropdown with a lot of options in the form of the **Delete** menu, as shown in *Figure 1.19*.

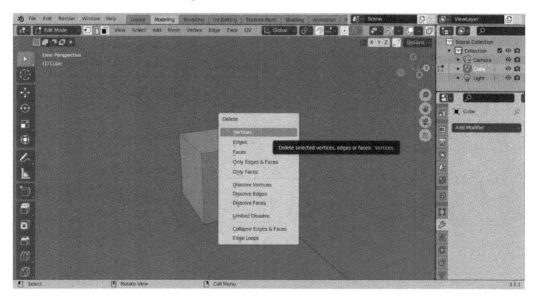

Figure 1.19 – The Delete menu

Let us understand the **Delete** menu in detail:

- Deleting **Vertices** will remove everything you have selected
- Deleting **Edges** will delete the edges you have selected and any faces connected to them
- Deleting **Faces** will delete faces and any edges that do not have any other faces connected to them to prevent floating edges
- Deleting **Only Edges & Faces** will delete faces and selected edges
- Deleting **Only Faces** will only delete faces without deleting any edges
- **Dissolve** will remove the selected geometry and create a face to prevent any holes

To undo a transform, press *Ctrl + Z*. To redo a transform that you undid, press *Ctrl + Shift + Z*.

In this section, we managed to add geometry while interacting directly with the mesh. There are also ways to change the mesh indirectly.

Manipulating a mesh using modifiers

Modeling operations are generally broken up into two general categories, destructive, and non-destructive. Their names are a bit misleading because you can both add and remove geometry by doing both types of operations. The difference is how they go about doing it. All of the modeling methods introduced so far have been destructive. **Destructive modeling** is when you perform an operation that cannot be tweaked or changed after finalizing the transform. The only way to undo a destructive transform is by using *Ctrl + Z*. **Non-destructive** transforms can be changed or completely removed without affecting the base mesh. These transforms affect the mesh indirectly, and are usually used outside of **Edit Mode**.

There are plenty of ways to affect the mesh indirectly by performing an operation on it outside of **Edit Mode**. One of the most common ways of doing this is through the **Modifier Properties** tab in the **Properties Editor Type** area. The **Modifier Properties** tab is used to manage modifiers. **Modifiers** are used to perform non-destructive operations on an object. The **Modifier Properties** tab can be seen in *Figure 1.20*.

Figure 1.20 – The Modifier Properties tab

The Mirror modifier

An example of one of these modifiers is the **Mirror modifier**. The **Mirror** modifier mirrors the object it is put on around the *X*, *Y*, or *Z* axis. To use a modifier on an object, follow these steps:

1. Make sure the object you want to add it to is selected.

2. In the **Modifier Properties** tab, press **Add Modifier** and navigate to the desired modifier – in this case, the **Mirror** modifier, as shown in *Figure 1.21*.

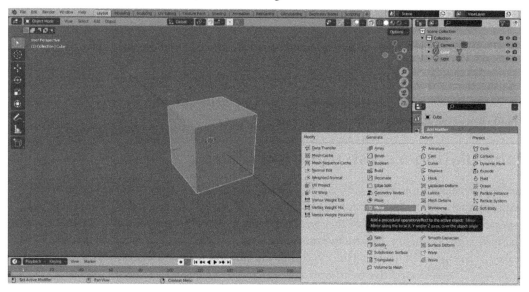

Figure 1.21 – The Mirror modifier in the Add Modifier tab

3. Finish adding the modifier by selecting it, and your modifier will appear in the **Modifier Properties** tab, as displayed in *Figure 1.22*.

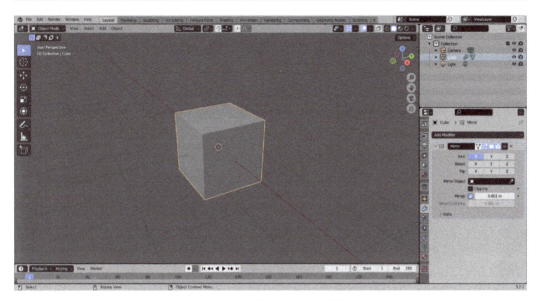

Figure 1.22 – The Mirror modifier

The **Mirror** modifier works by mirroring the mesh around an object's origin. The **object origin** is the little orange dot you see in **Object Mode** that indicates the object's orientation compared to the world's origin. *Figure 1.23* shows us a closer look at the origin.

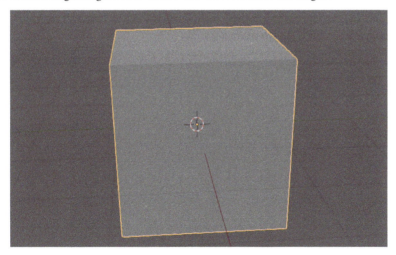

Figure 1.23 – The origin of an object

4. Pressing *N* to open the side panel shows the transforms of the object.
5. Translating the cube in **Object Mode** by pressing *G* and then the *LMB* to apply the translation will change these variables. This is captured in *Figure 1.24*.

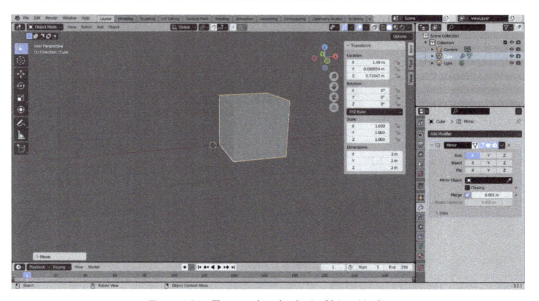

Figure 1.24 – The translated cube in Object Mode

The **Mirror** modifier will not be affected by a translation in **Object Mode** because translation in this mode also moves the origin, and the **Mirror** modifier mirrors around that.

6.   To see what the **Mirror** modifier is doing, we have to edit it in **Edit Mode** because the location of the vertices in this mode are all referenced to the origin of the object. If we translate the cube in **Edit Mode**, the result is shown in *Figure 1.25*.

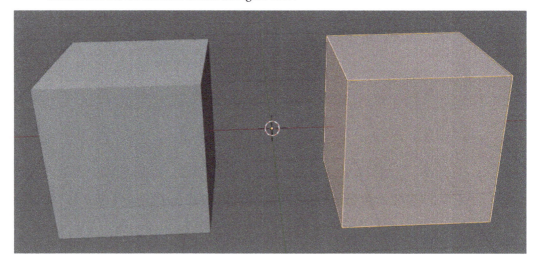

Figure 1.25 – The translating mesh in Edit Mode

Now we have a perfectly mirrored cube on either side of our origin. Next, we will look at how to connect the two mirrored halves. To connect the cubes together, follow these steps:

1. Start by selecting the face facing the origin and deleting just the face. This is to ensure that we do not have any extra faces inside of the mesh.

2. Next, on the **Mirror** modifier, make sure **Clipping** is turned on.

3. Finally, select the ring of vertices closest to the origin and translate them to the middle, while locking it on the *X* axis by pressing *G* and then *X*. The result should look like *Figure 1.26*.

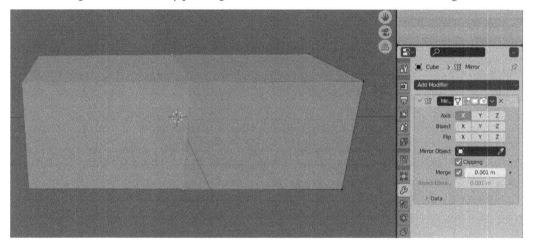

Figure 1.26 – Translating in Edit Mode with clipping enabled

The Subdivision Surface modifier

Another common modifier we are going to use is called the **Subdivision Surface** modifier, or *SubD* for short. The SubD modifier subdivides the mesh and also performs a smoothing operation on the subdivision. The levels of subdivisions can be adjusted for the viewport and the render can be adjusted separately. Modifiers can also be used destructively. You can apply them by pressing the arrow at the top right of the modifier, pointed out in *Figure 1.27*.

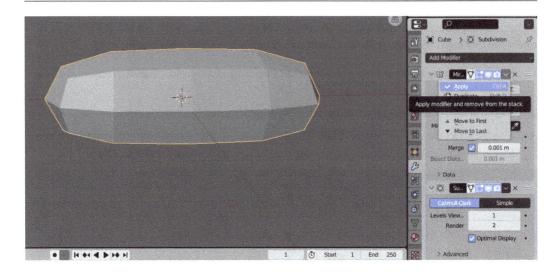

Figure 1.27 – Applying modifiers in the Modifier tab

Modifiers affect the mesh from the top of the modifier stack down. So, in this instance, the **Mirror** modifier is being calculated before the SubD modifier. It is usually best to apply modifiers from the top down to avoid any issues. You must be in **Object Mode** to apply modifiers. *Figure 1.28* shows the mesh in **Edit Mode** before applying modifiers, and *Figure 1.29* shows the mesh after applying them.

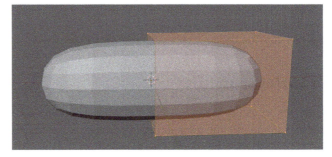

Figure 1.28 – The mesh in Edit Mode before applying the modifier

Figure 1.29 – The mesh in Edit Mode after applying the modifier

Before applying, the mesh does not change because it is non-destructive. After applying it, it becomes destructive and allows you to modify the mesh directly again. Now that we have an understanding of the **Mirror** modifier, the next modifier we will use will help us transfer the shape of one mesh to another.

The Shrinkwrap modifier

There are also modifiers that affect the position of the vertices non-destructively. An example would be the **Shrinkwrap modifier**. The **Shrinkwrap** modifier projects the vertices from one object onto another object. As an example, we are going to shrink-wrap a cube onto a sphere:

1. Go into **Object Mode** and add a cube by pressing *Shift + A*.

2. Hover over the mesh.

3. Select the cube shown in *Figure 1.30*.

4. Repeat *1* to *3*, only this time, select the UV sphere.

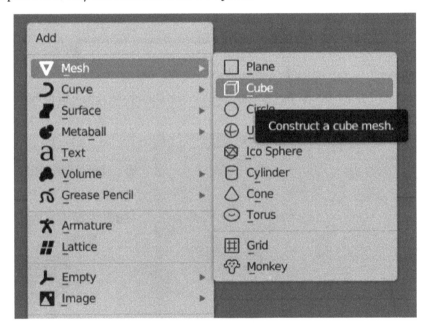

Figure 1.30 – Adding a primitive cube

After both objects have been added, it should look something like *Figure 1.31*.

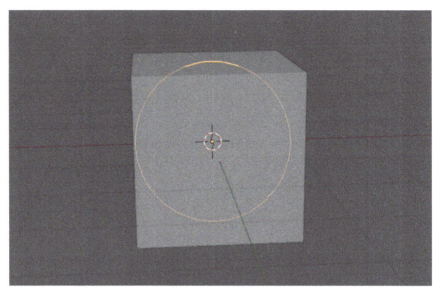

Figure 1.31 – The sphere and cube added on top of each other

5.   After this point, we can select the cube and add our **Shrinkwrap** modifier.

6.   Select the target option on the **Shrinkwrap** modifier and choose our sphere from the dropdown illustrated in *Figure 1.32*.

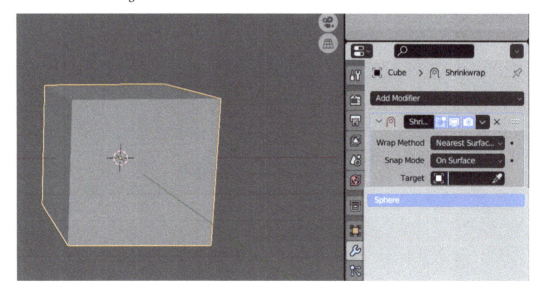

Figure 1.32 – The Object selection tab

As soon as this is done, the eight vertices of the cube will be cast onto the surface of the sphere. If we want the cube to conform to the sphere, we will have to give it more geometry to work with. That is where we can use the SubD modifier.

Remember, the modifier stack is calculated from the top down, and we want to add the extra geometry, so we need to put the SubD above the **Shrinkwrap** modifier.

This can be done by grabbing the top right of the modifier and dragging it above the other modifier. After that, just add another level of subdivision to the viewport, and it should look like *Figure 1.33*.

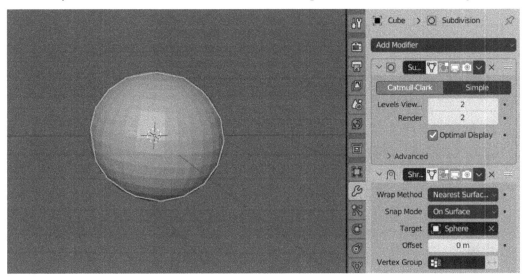

Figure 1.33 – The cube shrink-wrapped to the sphere

Summary

In this chapter, you learned about meshes, objects, transforms using hotkeys, the difference between destructive and non-destructive transforms, and modifiers.

Now that we know all about our mesh and how to manipulate it, we should be ready to hop into some real topology. In the next chapter, we will be using many of these transforming techniques to illustrate how the topology of a mesh is supposed to behave.

2

The Fundamentals of Topology

In the previous chapter, we learned about the Blender UI and how we work on objects within Blender using **Edit Mode** and **Object Mode**. In this chapter, we will start on the basics of 3D modeling by forming an understanding of what topology is. **Topology** is a method that is the building block of all modeling and determines the composition of the faces and edges in a mesh. It determines the shape of an object and how many triangles it has. It determines the way that it will deform, and how materials are applied to it. Without good topology, rigging, UV unwrapping, and soft body simulations are all made far more difficult. We will understand these terms and learn why they are important in *Chapter 3* and *Chapter 4*.

Regardless, first, we must build a basic understanding of topology. In this chapter, we will learn how to manipulate topology to achieve the shape we want. Then, we will learn how to merge separate topologies into each other. Finally, we will learn how to identify the different ways topology will flow for different shapes.

In this chapter, we will be learning about the following subjects:

- Understanding good topology using grids
- Understanding the three rules of topology
- How should grids intersect?
- Identifying grids on a complex shape

Understanding good topology using grids

Good topology is a mesh that deforms the way you want and makes the shape you want with the fewest number of triangles. The easiest way to get clean topology is by using a quad-based workflow. A **quad** is a face that contains four edges and four vertices. Although it is considered a single face, it is technically two triangles combined. This can be seen in *Figure 2.1*. Because you can manipulate the vertices individually, it must contain two triangles.

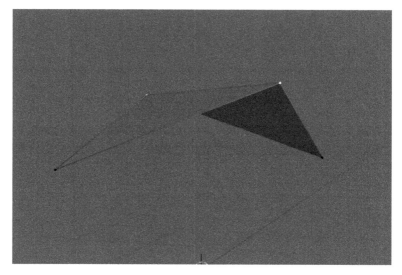

Figure 2.1 – Quad made up of two triangles

If you were to take multiple quads and string them together to form a plane, you would get a grid as shown in *Figure 2.2*.

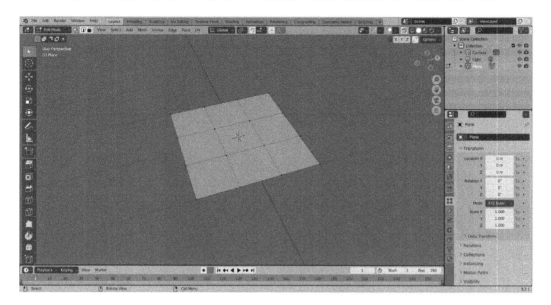

Figure 2.2 – A grid (the perfect topology for a plane)

If what you were planning to make was a plane, this would be an example of perfect topology. Unfortunately, objects and characters are often a bit more complex. Thankfully, there are a few useful tools we can use to help us check out topology using the grid as our example.

The most useful tool we have to check what our topology is doing is the *Loop Cut* tool. The *Loop Cut* tool adds a loop of edges through a row of connected faces. To use this tool, go through the following steps:

1. To begin a loop-cutting operation, press *Ctrl* + *R*.

2. After entering the operation, you can hover over a face and a line will appear along a loop of faces designating where the loop cut will take place – this is illustrated in *Figure 2.3*.

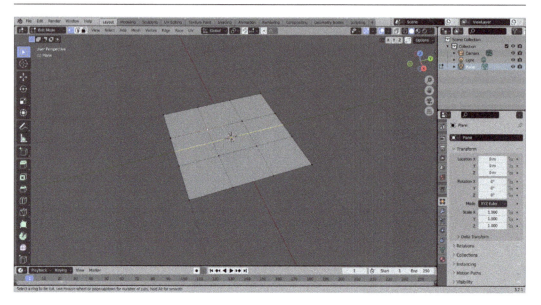

Figure 2.3 – Loop cut going through the plane

3.   To finalize the loop cut, press *LMB*, and to cancel it, press *RMB*. The finalized loop is shown in *Figure 2.4*.

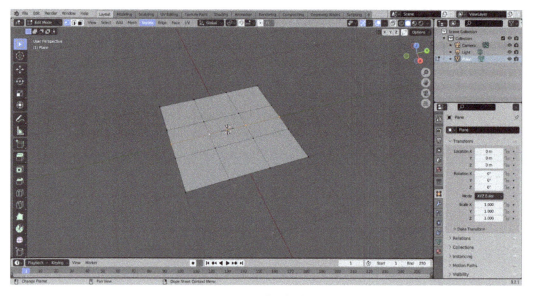

Figure 2.4 – Finalized edge loop

4. To add more loops to the cut, scroll up on the mouse wheel. To add fewer, scroll down, or you could press the number of loops you wanted on the number line and number pad.

For the purpose of checking the topology, we will mostly be using the guiding loop to see where our loops are going, so we will cancel the actual loop cuts. If we run a loop cut along one of the rows of faces shown in *Figure 2.5*, a few things are apparent.

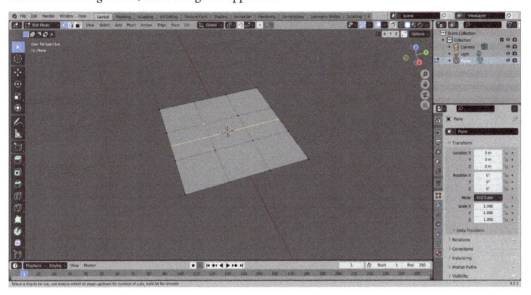

Figure 2.5 – Loop cut for checking topology

- First, the loop cut is actually going through the faces, showing us that our grid is properly made of quads

- The second is that the line is going straight through the mesh, terminating on either end into the void

- We can repeat this loop on every part of our grid, from every direction, and this is still true

However, one thing may cause issues here if you have started an extrusion on one of the points and then canceled it by pressing *RMB*. Because *RMB* does not remove the geometry created from the extrusion, you may be left with overlapping edges. In *Figure 2.6*, you can see what this might look like. Although it appears as though it is a clean grid, there is a duplicate edge blocking the loop cut.

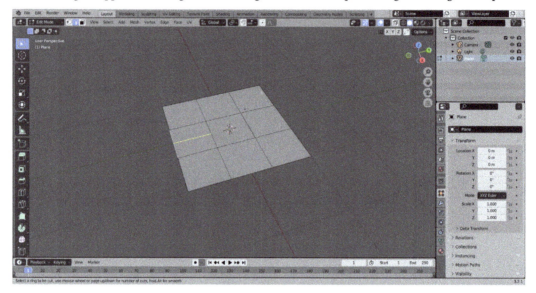

Figure 2.6 – Broken loop cut caused by an extrusion

This issue is usually solved by merging by distance. **Merging by distance**, also called **Remove Doubles**, merges the vertices of a selected mesh by the desired distance. To perform a *Merge by Distance* operation, see the following:

1. Start by selecting the mesh with *A*.

2. Press *M* to open the **Merge** tab.

3. Then, select **By Distance** from the dropdown.

You can see the menu in *Figure 2.7*.

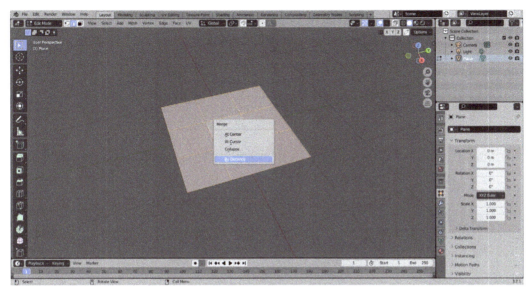

Figure 2.7 – Merge menu

Another thing that can stop a loop from going through is a face flipped the wrong way. A face's direction is determined by its normal. **Normals** are used to determine the direction in which the light will bounce off of a surface. A face has a front face and a back face determined by this normal. You can see a face normal direction by applying some settings in the **Viewport Overlays** tab shown in *Figure 2.8*. These settings are different in **Object Mode** and **Edit Mode**. For determining whether a face is flipped or not, check the **Face Orientation** checkbox in the **Object Mode** overlays tab. This will make all of the front faces blue, and all of the back faces red.

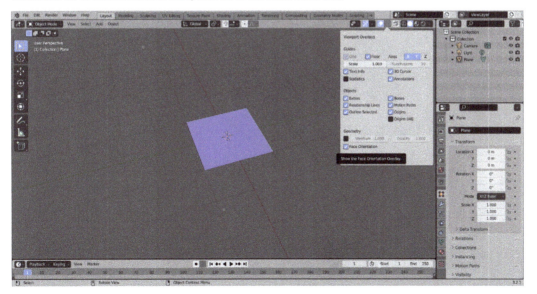

Figure 2.8 – Face Orientation checkbox

You can see the face that is flipped on this mesh in *Figure 2.9* with **Face Orientation** enabled.

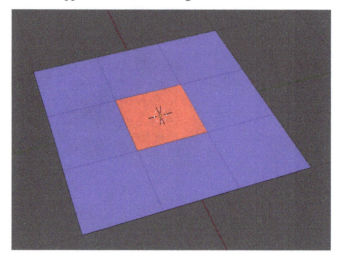

Figure 2.9 – Face Orientation revealing a flipped face

To change the orientation of a face, there are two options. You can press *Alt + N* to open up the **Normals** tab displayed in *Figure 2.10*.

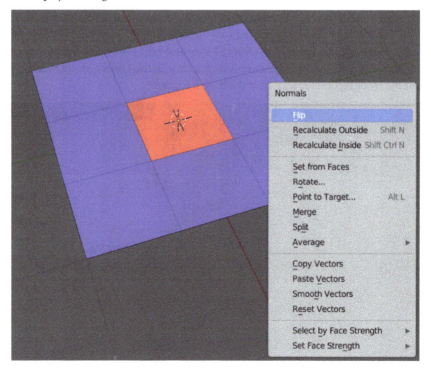

Figure 2.10 – Normals menu

In this tab, you can flip a selection of faces by selecting **Flip** at the top, recalculate faces on the outside of the mesh, or recalculate them on the inside. The most common option used here is **Recalculate Outside**. This option will fix flipped faces most of the time. The hotkey to recalculate the normals on the outside is *Shift + N*. In *Figure 2.11*, the face is fixed, and the loop goes through all of the faces as desired.

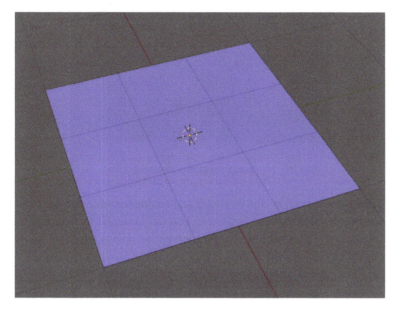

Figure 2.11 – Fixed normals

By now, you should have a basic understanding of how a grid works, and how we can check the topology of that grid to judge whether it constitutes good topology. However, to work on more complex objects and manipulate grids to improve the topology, we need a stronger understanding of the three rules of topology.

Understanding the three rules of topology

Now that we have a basic understanding of how a grid works, we can apply that understanding to a few more shapes. To do this, first, we need to understand the three conditions of good topology.

Rule 1 – an edge loop must terminate into the void or into itself

The first condition is that an edge loop needs to either terminate into the void or directly into itself. A cube is a perfect example of this. Like a plane, a default cube has the perfect topology for a cube. If you run a loop cut along any of the sides, all of the loops meet themselves all the way around. This is illustrated in *Figure 2.12*.

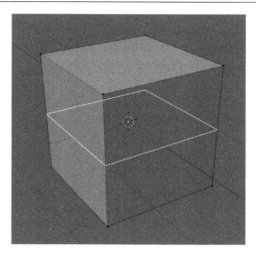

Figure 2.12 – Loop cut going around a cube

A cube is just six grids all joined end to end, with each grid having the same number of faces on them. If you were to isolate one of the sides, as in *Figure 2.13*, you would be able to run loops through it like any other grid. This is the core idea behind this approach to the topology. Everything can be broken down into simple grids.

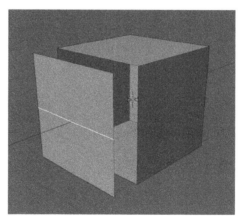

Figure 2.13 – Loop cut going through one face of the cube

This principle is well illustrated by our next exercise. If we take our cube and run it through a few operations, we can turn it into a cylinder with the same topology.

Here are the steps we need to take to turn our cube into a cylinder:

1.   First, we go into **Edit Mode** and select our top and bottom face.

2.   Next, we delete the faces by pressing *X* and selecting **Faces**.

3. Then, we go back to our modifier tab and add a subdivision surface.

4. Increase the number of viewport subdivisions to 2.

5. After that, we go into **Object Mode** and apply the modifier as shown in *Figure 2.14*.

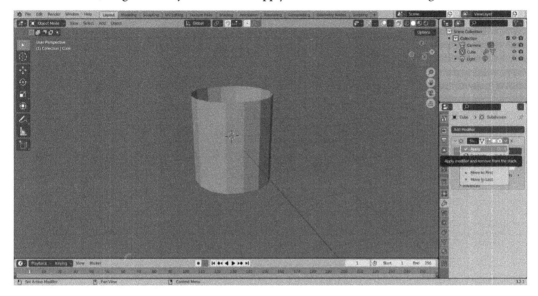

Figure 2.14 – Cylinder made from a cube

Figure 2.15 shows the mesh after the subdivision surface was applied. Now, we can do the same checks with the **Loop Cut** tool as we did previously. If we run loop cuts from all directions, the loop either wraps into itself or goes into the void. One thing to note here is that even though we used a subdivision modifier, it did not actually change the flow of our topology. It just added mode geometry and changed the shape.

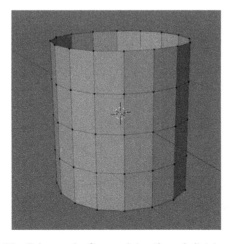

Figure 2.15 – Cube mesh after applying the subdivision modifier

When working with quads, the subdivision surface will not change your topology, and this is very important to take on board now because if you end up applying your subdivision and then notice a mistake afterward, it is much more difficult to fix.

Rule 2 – loops must not intersect themselves

Our next rule is that loops cannot intersect themselves. This might seem contradictory to the last rule, which is that loops can wrap back into themselves, but if you look at *Figure 2.16*, you will see how this second rule is different from the first one.

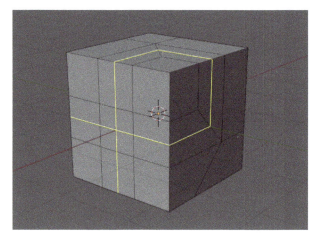

Figure 2.16 – Loop cut intersecting itself

Instead of meeting itself to form a coherent loop, it cuts straight through itself, perpendicular to the part of the loop it intersects. If we flatten the mesh into a flat plane, as in *Figure 2.17*, it is easier to see how the grid was changed to form this issue. It is as though we took one of the sides of the grid and attached a side perpendicular to it.

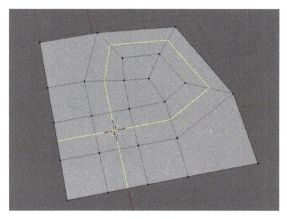

Figure 2.17 – Flattened version of an intersecting loop cut

Rule 3 – loops must not spiral down a mesh

The last rule is that we do not want these loops to spiral down a mesh. This happens when a loop is not joined back into itself and is joined to a different parallel face. This is shown in *Figure 2.18*. This might seem acceptable because it does not cut through itself and terminates in the void, but this can cause serious issues when trying to deform it. When performing later operations such as UV unwrapping, or any sort of deformation, this is a huge pain to deal with.

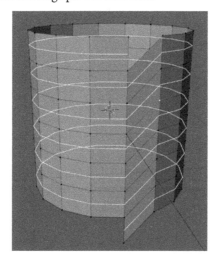

Figure 2.18 – Spiraling loop cut

This also leaves faces at the top and bottom that are almost impossible to terminate without breaking the other rules and makes operations that need to run along the corner of the mesh, such as bevels, impossible. You can see one of these faces highlighted in *Figure 2.19*.

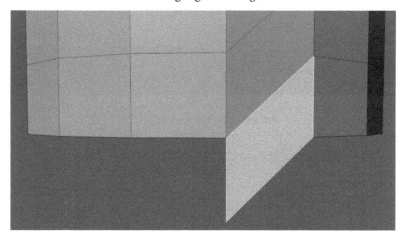

Figure 2.19 – Extra face at the bottom of the spiral

These rules need to be applied to almost every mesh we will be working with, so before moving on, we should summarize what these three rules and good topology imply. You should keep all of these in mind, especially as we move into the next section, where we will be intersecting grids:

- All of the faces in a mesh are quads

- All of the normals on the mesh are facing in the right direction

- All of the loops terminate into themselves or the void

- None of the loops overlap themselves

- None of the loops spiral around the mesh

We now know how the three rules apply to a grid-like topology, so we will look at how to apply these rules to intersecting grids next.

How should grids intersect?

Our first example of intersecting grids will start with a simple grid, which you can see in *Figure 2.20*.

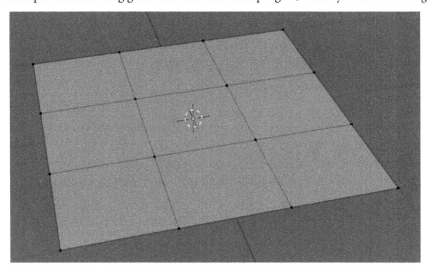

Figure 2.20 – Normal grid

To create a grid, as in the preceding figure, see the following:

1. Go into **Edit Mode**.

2. Perform a loop cut on the plane by pressing *Ctrl + R*.

3. Scroll up once on the mouse wheel so that you have two loop cuts, resulting in *Figure 2.21*.

Figure 2.21 – Plane with two loop cuts

4. Repeat this double loop cut perpendicular to the first cuts.

Now that we have a matching grid, we can tweak it to help illustrate an intersecting grid. We are going to extrude a face from the center of the plane to see how it affects the topology:

5. First, select the center face.

6. Press *E* to extrude the face.

7. Move the mouse so that the selected face is pulled away from the rest of the faces to make an extrusion.

8. Finalize the transform by pressing *LMB*.

After finalizing the transform, it should look something like *Figure 2.22*.

Figure 2.22 – The plane with the extruded face

Now, we can run through our tests to see whether this topology still follows all of our topology rules:

- If we run a loop across the extrusion as in *Figure 2.23*, you can see that it terminates into the void, doesn't intersect itself, and does not do any weird spiraling.

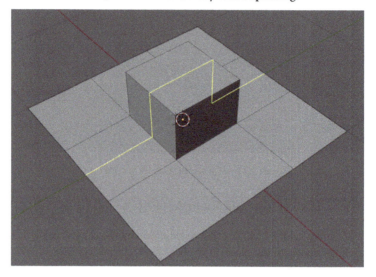

Figure 2.23 – Loop cut checking the topology across the extrusion

- If we run a loop around the extrusion like in *Figure 2.24*, you can see that it terminates into itself to form a closed loop. It does not intersect itself, and it does not spiral around the mesh.

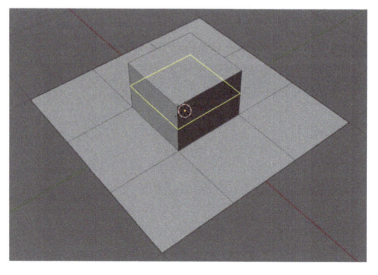

Figure 2.24 – Loop cut for checking the topology around the extrusion

Now that we have confirmed that the extrusion we did follows all of our topology rules, we can take a look at some of the defining features of this shape. In terms of topology, what we have here is a cube that has intersected a grid. We will start by examining the vertex highlighted in *Figure 2.25*.

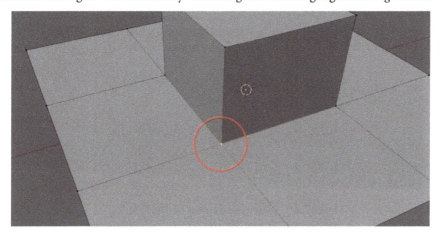

Figure 2.25 – Highlighted vertex circled in red

In the preceding figure, a vertex is highlighted at the corner of the intersection between the cube and the plane. This corner is called a pole. A **pole** is a vertex that has more than four edges intersecting at the vertex. As you can see in our example, this pole has five vertices. For quad-based topology, you usually do not want more than five edges on your poles. If you take a look at a normal vertex on a grid, they have four edges connecting them. If we flatten our shape back into a plane, as shown in *Figure 2.26*, we get a new perspective of the same topology.

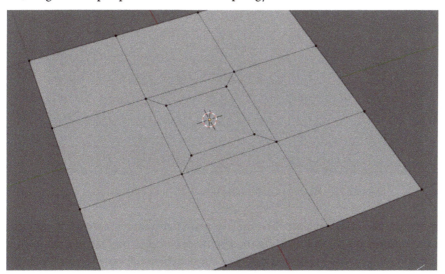

Figure 2.26 – Flattened version of the extruded topology

If we run our loop cuts over this model to check out the topology, you can see that it also follows our rules. It terminates into the void, does not intersect itself, and does not spiral. This has the same topology as our intersecting cube, just a different shape. With this new perspective, we can look at that corner pole again in *Figure 2.27*.

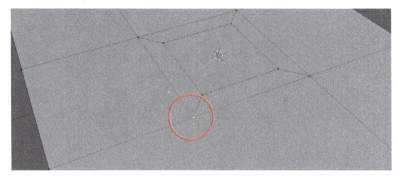

Figure 2.27 – Highlighted vertex circled in red

Now, it is a lot easier to see where this fifth edge on the vertex is coming from. It is like we shrunk the middle face and added four new faces. Thinking back to what we learned about faces earlier, you need four edges per face. This means we need another four edges added to make sure each face has four edges. In *Figure 2.28*, you can see the old edges marked in red, and the new edges used to make the new faces marked in blue.

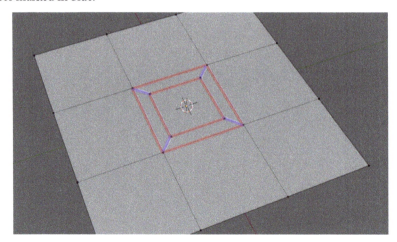

Figure 2.28 – Flattened extruded plane with highlighted edges

Everywhere that the two grids intersect, there are five-point poles at their corners. It can be difficult to tell where the different intersecting grids are on a more complex mesh. An example of how to identify these grids on a more complicated model will be helpful to understand how topology works in a practical situation.

Identifying grids on a complex shape

We are going to be using this object shaped like a pair of pants in *Figure 2.29* as our practice shape.

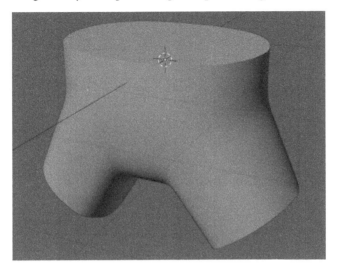

Figure 2.29 – Practice shape

In our example, we will be using a tool called snapping. **Snapping** snaps your selection to an edge, face, vertex, or volume. The example uses snapping to faces. Snapping to faces can be enabled by pressing the magnet at the top of the viewport, and the options can be shown by pressing the arrow next to the magnet. The settings can be seen in *Figure 2.30*.

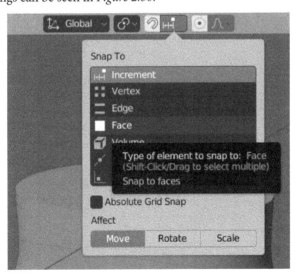

Figure 2.30 – Snap To menu

To start, we will identify the defining shapes of the mesh. All we are looking for are the specific areas that we know we want the edges to line up with. These areas are highlighted in *Figure 2.31*.

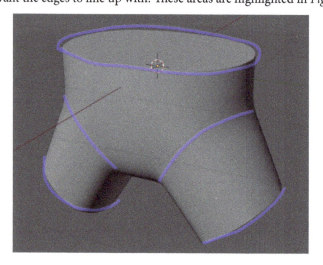

Figure 2.31 – Areas to line our vertices to

The next steps are going to give you an overview of the general process we are going to use when applying our topology rules to a model. Do not be afraid if there are specific steps that you do not understand at this point, as we will focus on this process in depth later in the book:

1.   First, we lay out vertices along these specific areas of detail, spacing them out so that it forms the shape naturally, as in *Figure 2.32*.

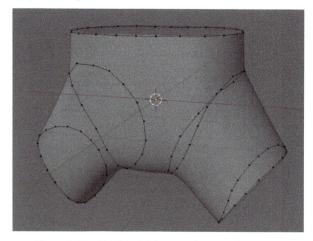

Figure 2.32 – Vertices laid out in the specified areas

2.   Use a **Mirror** modifier and mirror on the *x*- and *y*-axes.

3. Enable **Clamping** on the **Mirror** modifier.

4. Then, you select the loops individually, extrude them out in both directions by pressing *E*, and finalize the transform by pressing *LMB*. After this step, it should look something like *Figure 2.33*.

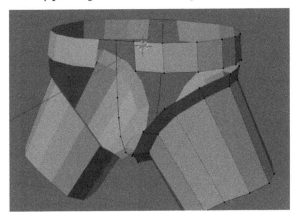

Figure 2.33 – Extruded vertices

5. Next, move the vertices to fit the shape as necessary by selecting the vertex you want to move and pressing *G*. Make sure to pull the vertices around the hip down to the bottom of the pant leg. After repositioning, it should look like *Figure 2.34*.

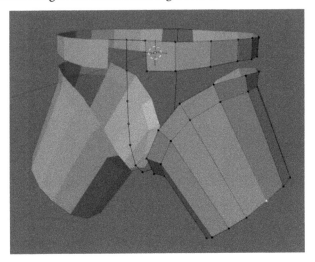

Figure 2.34 – Mesh with vertices snapped in place

In the preceding figure, you can see the separate grids that we need to connect. If you look at all of the loops on this mesh, they all follow our rules. The tricky part is joining them so that the loops still follows all of our rules. We will start with some of the easy parts.

6. Around the sides, we can create faces by selecting edges that are lined up with each other and pressing *F*. This is shown in *Figure 2.35*.

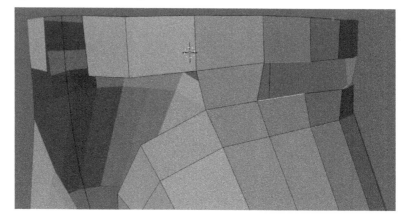

Figure 2.35 – Face made between edges that line up

7. The same thing can be done along the bottom joint of the pants. The result is illustrated in *Figure 2.36*.

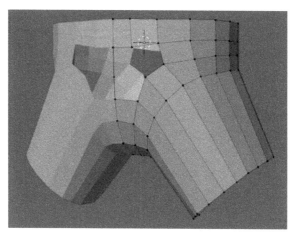

Figure 2.36 – The result after joining all of the edges

8. If you have too many edges on either side, just select the row of edges that is causing the problem, and dissolve it by pressing *X* and then selecting **Dissolve Edges**. Make sure to select and dissolve the whole edge loop.

We can check our mesh now with some loop cuts, and it should still follow all of our rules. It terminates into the void, does not intersect itself, and does not spiral.

Now, we need to look at the trickier parts. In *Figure 2.37*, you can see the areas in question.

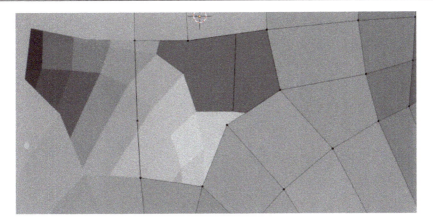

Figure 2.37 – Tricky areas

Unlike the previous connections, there is no straight path for these loops to follow; they need to connect at an angle. Therefore, in this case, we are just going to choose the path that seems more natural for each face. After connecting the faces, your model should look something like *Figure 2.38*.

Figure 2.38 – Final mesh

Here, one loop continues in the direction of our previous loops, and the one next to it loops across to the other side. You will notice that right where that split happens, we now have a pole, a vertex with five edges coming off of it. If you recall those initial grids we defined in *Figure 2.33*, there are two separate grids intersecting each other. These poles are formed at the corners of those intersecting grids.

If we flatten out our pants, it is much easier to see what is happening. *Figure 2.39* shows it flattened, but with the same topology.

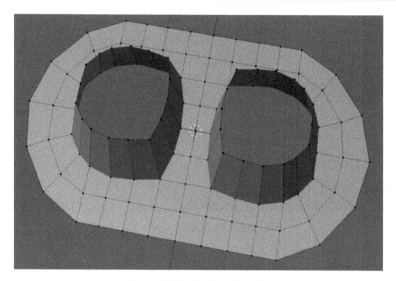

Figure 2.39 – Flattened mesh

You will notice right away that the middle looks like a normal grid, with two intersecting grids to form the legs. To identify separate grids intersecting, just look for the poles. At every intersection with another grid, there is at least one pole. Then, all we need to do is run our loop cuts along the mesh to identify where the loops are flowing to see what the grid is doing.

Summary

In this chapter, we learned how to use loop cuts to check out topology. We also learned about important principles and rules of good topology, such as that all of the faces in a mesh should be quads, all of the normals on the mesh need to face the same direction, all of the loops on a grid need to terminate into themselves or the void, none of the loops can overlap themselves, and loops cannot spiral around the mesh. We also learned that intersections between grids can be identified by poles.

Now that we have learned about all of these rules, we can start to apply them, and learn why we should follow these rules. In the next chapter, we will look at deforming meshes and how we should position our topology for a good deformation.

3

Deforming Topology

In *Chapter 2*, we were introduced to the quad-based grids that make up models. One of the important functions topology serves is determining how a model is going to deform when using an armature, or doing a physics simulation. All of the rules that we learned are going to help us ensure that the model deforms the way that we want it to.

To get more comfortable with topology rules and how they relate to the deformations, we will be introduced to some standard joints and deforming geometries. We will work with bending joints, compressing and stretching meshes, and folding cloth using simulations.

The three rules discussed in this chapter are what I have discovered necessary to apply while working with 3D models that need to be bent, stretched, or twisted. In this chapter, you will understand both what these rules are and why they are important to apply.

In this chapter, we will be learning about the following subjects:

- Applying the bending and stretching deformation rule
- Applying the twisting deformation rule
- Applying the intersecting grids deformation rule
- Topology for cloth simulations

Applying the bending and stretching deformation rule

In *Figure 3.1*, you can see an elbow joint. This is one of the most common joints modeled on a mesh being deformed by an armature. **Deformation** refers to the bending or stretching of a mesh and is used to describe any shifting of the mesh after the mesh is finished.

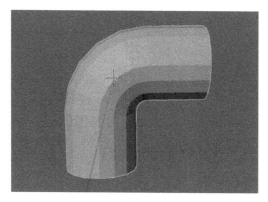

Figure 3.1 – An elbow joint

Our **first deformation rule** is that the edges of our grid need to be parallel to the axis of deformation. That is a bit of a mouthful, so we are going to break it down with a few examples. In *Figure 3.2*, we have a plane that has been bent around the *y*-axis.

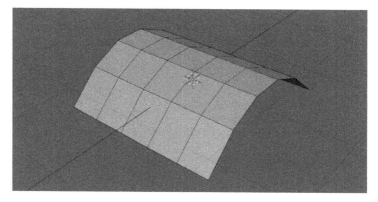

Figure 3.2 – A plane bent around the y-axis

Go through the following steps to make this shape:

1. First, start by pressing *Shift + A*.

2. Add a plane from the mesh dropdown.

3. Then, go into **Edit Mode**.

4. Click on **RMB**, and select **Subdivide**.

5. In the bottom left of the Viewport, the **Subdivide** tab will appear – increase the number here until you are satisfied with the number of faces.

6. Select each of the edges and move them individually.

7. To move them along a specific axis, press *G* and then *X*, *Y*, or *Z* until it roughly matches *Figure 3.2*.

Now that we know how a plane is supposed to deform, we are going to look at how it is not supposed to deform. In *Figure 3.3*, you can see a lot of jagged corners have formed because we bent the plane the wrong way.

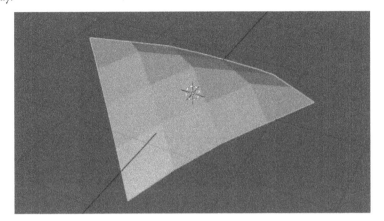

Figure 3.3 – Jagged corners resulting from bending the plane the wrong way

Because of the way the quads are positioned, we do not have a lot of control over which way the triangles are deforming. When we compare that to our mesh with the proper deformation in *Figure 3.4*, it is easy to tell the difference between the two.

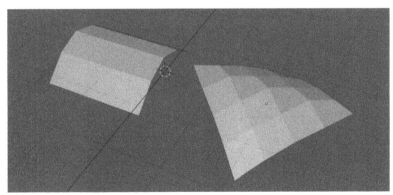

Figure 3.4 – Comparison between proper deformation (left) and improper deformation (right)

Next, we are going to look at a more commonly deformed shape, a cylinder. Remember, a cylinder has the same topology as a plane, just wrapped back into itself, so just like our plane, we need to bend it along the edges or make sure that our edges match up with our axis of deformation. If we were to bend it along the *x*-axis, it would look something like *Figure 3.5*. Notice how all of the loops of vertices wrapping around the mesh are in line with the *x*-axis. This holds true with any deformation axis you choose. It could be the *x*, *y*, or *z*-axis – or a random direction. Either way, the edges across the model need to be parallel to that axis of deformation.

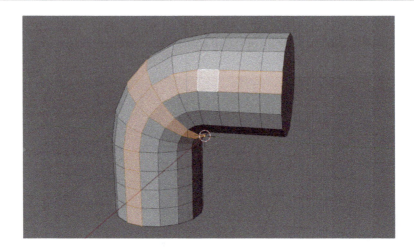

Figure 3.5 – A properly deformed cylinder

Figure 3.6 shows a deformed cylinder that does not follow this rule.

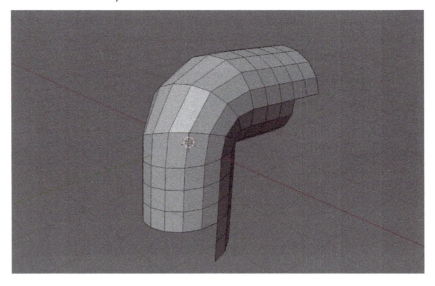

Figure 3.6 – An improperly deformed cylinder

To get a better look at what is going on, we should reset this cylinder back to its original form. You can see this in *Figure 3.7*.

Figure 3.7 – A spiraling topology that can lead to improper deformation

This shape should feel familiar because we saw it in *Figure 2.18* in *Chapter 2, The Fundamentals of Topology*. It was an example of a topology that breaks one of our topology rules: loops must not spiral down a mesh.

Figure 3.7 is an example of spiraling topology, and when trying to deform a joint, it can cause serious issues as shown before. Any time you have spiraling like this, you just have to isolate where it began and figure out how to connect it back into itself properly. In this case, you would have to shift the faces as illustrated in *Figure 3.8*.

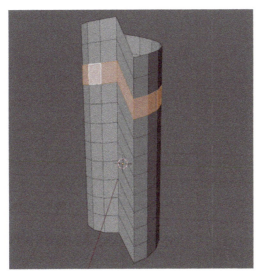

Figure 3.8 – Shifting the face of the spiraling topology to correct the mismatch

These spirals are usually caused by a mismatch, so finding that mismatch usually fixes the issue. Looking back at our regular cylinder deform, if you focus on the bend like in *Figure 3.9*, you can see it still looks a little rough.

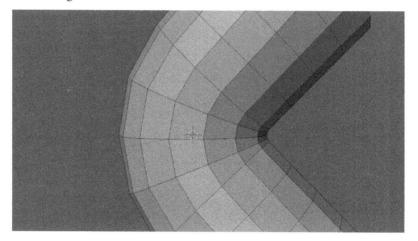

Figure 3.9 – Zooming in on the deformed pipe's bend

The faces on the back look stretched, and the faces in the middle are squished.

You can try to fix this by adding more loops before deforming the cylinder, but it then causes the loops in the crease of the joint to become even denser. You can see it deformed again in *Figure 3.10*.

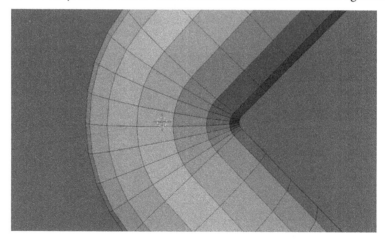

Figure 3.10 – The bend in the deformed pipe with more loops added

Likewise, if we remove too many from the inside, the back will stretch too much. What we want to do ideally is have more geometry on the back, and less geometry on the inside. Before we try to do this on an elbow joint, we should take a closer at stretching and compressing topology.

Our closer look at stretching and compressing, again, starts with a grid. If we look at *Figure 3.11*, we can see how a mesh is supposed to stretch and compress. You can see the left side stretches out, in the middle, it becomes a square again, and on the right, it is squished together.

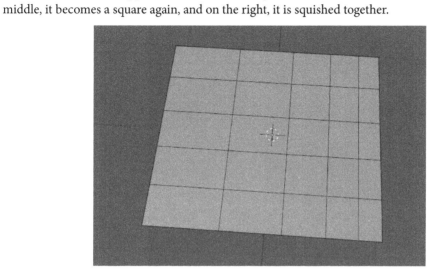

Figure 3.11 – A stretched and compressed grid with the proper deformation

An example of a deformation we do not want can be seen in *Figure 3.12*.

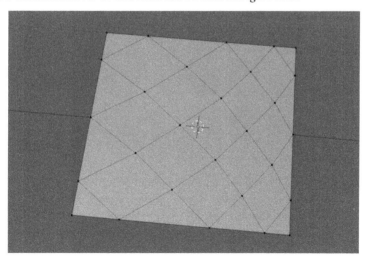

Figure 3.12 – A stretched and compressed grid with improper deformation

You can see right away the preceding grid looks messy, and instead of compressing or stretching into rectangles, these rectangles stretch out across the corners and create diamond shapes. While it may not look bad while the plane is flat, if we start to bend it along the x or y-axis, it looks very rough. *Figure 3.13* shows the two side by side.

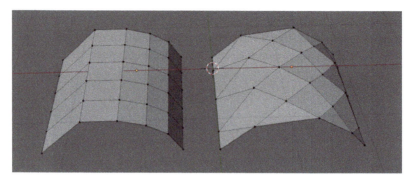

Figure 3.13 – Comparison of a properly deformed bent plane (left)
and an improperly deformed bent plane (right)

And this is just a plane. The deformations get even worse when we move to more complex shapes. If we make a loop out of our plane, we can make something similar to an eyelid or mouth. This is a common shape in topology to manage something that needs to be able to stretch or compress radially. You can see the loops in *Figure 3.14*. It also shows the directions in which the topology should be able to stretch.

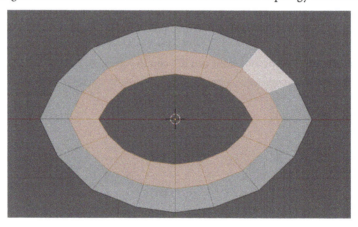

Figure 3.14 – A properly deformed plane looped to create a mouth or eyelid

If we deform our loop like an eyelid as in *Figure 3.15*, you can see that the faces that make up the topology retain more or less the same proportions, and the edges stay in line with the deformation.

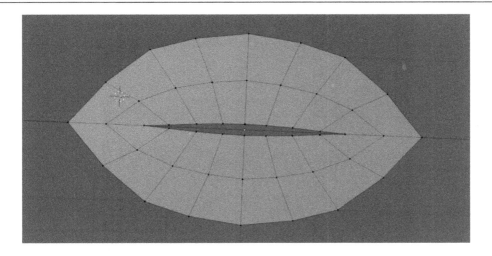

Figure 3.15 – A properly deformed plane looped to create an eyelid or mouth

As soon as you introduce lines that are not perpendicular to the stretching geometry, you get issues.

Identifying poor topology on a cylinder

Now that we know how poor topology might react to stretching on a plane, we can look at how it will affect a cylinder. You can see such a cylinder in *Figure 3.16*.

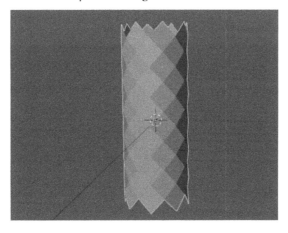

Figure 3.16 – A cylinder with poor topology

Immediately, it is easy to tell that something is off. The profile of the cylinder even looks weird. If we try to bend this model, it will result in a truly terrible deformation, as evident in this dreadful elbow joint in *Figure 3.17*.

Figure 3.17 – A bent cylinder with poor topology

In the preceding figure, the back is stretching much more than the previous elbow joint did, and the crease of the elbow is just as bad – just scrunched instead.

Fixing the topology on an elbow joint

Alright – we should take a closer look at that elbow joint we made earlier using a normal cylinder. We can see how it deformed again by looking back at *Figure 3.5*.

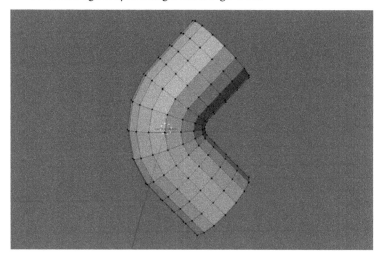

Figure 3.18 – A closer look at the elbow joint of a normal cylinder

We should try to see exactly where the mesh is stretching and compressing so that we can adjust the parts that we need to. Looking at *Figure 3.19*, we can see the specific areas we are having trouble with. On the outside of the joint, the faces are stretched out, and on the inside, the faces are scrunched.

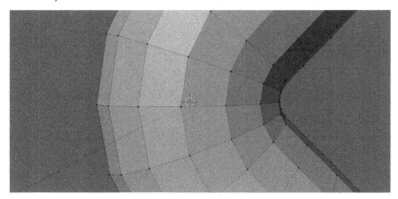

Figure 3.19 – A closer look at the faces on the bend

An important thing to notice is the faces in between them. While the outside and inside of the joint deform a lot, the middle parts only deform a little bit. This is an important thing to keep in mind because it means we do not have to worry as much about these areas.

Then, the question we need to ask now is how do we actually manipulate the topology? Again, it is time for us to look at our grid. You can see our friendly grid in *Figure 3.20*.

Figure 3.20 – A 4x4 grid

What we are going to focus on first is adding more geometry to the back of the joint. One of the easiest and most consistent ways to add more geometry and have it follow our topology rules is to perform an extrusion using the following steps:

1. First, select the inner faces.
2. Then, press *E* to extrude the mesh.

3. Then, press *S* to scale.

4. Move the mouse to the desired scale.

5. And finally, use *LMB* to apply the transform.

When you are done extruding and scaling, it should look something like *Figure 3.21*. You will notice that the faces on the inside of the extrusion are much smaller than the ones surrounding it.

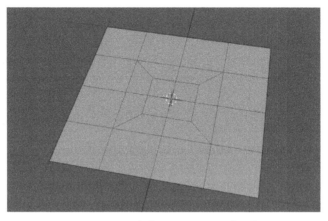

Figure 3.21 – Grid after extrusion and scaling

What allows the faces to shrink? If we take a closer look at it, we can see where the edges are forming this new geometry. Looking at *Figure 3.22*, we can see those poles on each of the corners of our extrusion.

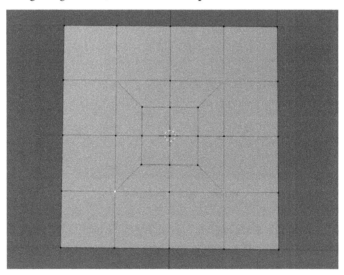

Figure 3.22 – A view of the poles on the corners of the extruded grid

These are ultimately where the new geometry is coming from. If you think back to the last chapter, you might remember that these poles indicate a corner where two grids are intersecting. That is essentially what is happening here. These faces are highlighted in *Figure 3.23* and are connecting the two grids that are not selected to add more geometry.

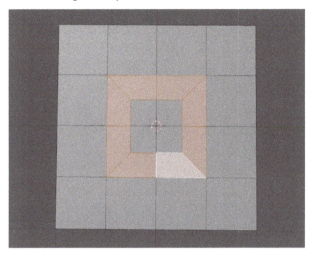

Figure 3.23 – The selected faces connecting the separate grids

This does cause a few problems right off the bat. Most notably, the faces connecting the two grids are warped quite a bit. We can help address this by moving the corners closer to the center point in the direction of the arrows in *Figure 3.24*.

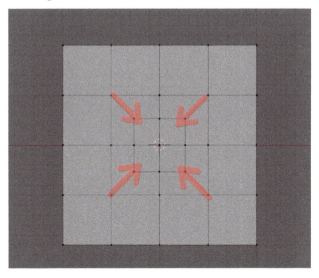

Figure 3.24 – Moving corners closer to the center

This will help to distribute the distortion more evenly along the surface of the mesh. After you have repositioned the corners of the mesh, it should look something like *Figure 3.25*.

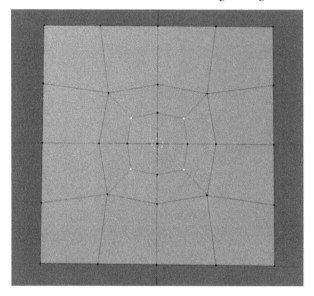

Figure 3.25 – Mesh with redistributed distortion

The second thing to note is that those sides joining the grids together are also still a lot longer than the smaller faces we just extruded. To start, we should try and compress one side of our new topology and see how it reacts. You can see it compressed in *Figure 3.26*.

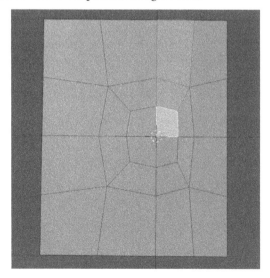

Figure 3.26 – Distorted topology with one side compressed

Overall, it looks pretty good. Almost all of the geometry is parallel or perpendicular to the direction in which we compressed it. The same thing can be said when stretching it as well. The only other major thing we need to worry about is how it bends, and this is where some issues start to pop up. If you remember, we are trying to stretch the denser part of the mesh, and compress the less dense areas.

You will notice that when we bend this mesh, some areas are denser than others, so it creates an uneven bend. This bend is visible in *Figure 3.27*.

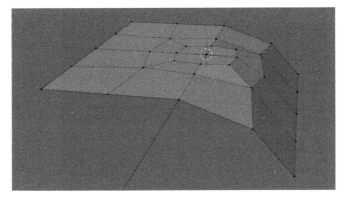

Figure 3.27 – A bent mesh with an uneven bend

The issue here is caused by the long faces on the edges, and the shorter ones on the inside. The difference in shape is causing a difference in the deformation along the face. Remember, we are not actually applying this to a plane. We are applying it to a cylinder that is only deforming in one direction. To start, we are going to add our geometry to our cylinder. To do this, we are going to select the faces at the back of the joint and extrude them as we did on the plane. After extruding it and moving the vertices, it should look like *Figure 3.28*.

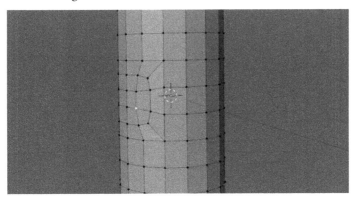

Figure 3.28 – Cylinder with extruded faces like the plane

The selection for the extrusion should be about halfway around the arm, with at least one row of faces below and above the loop of edges that will form our crease.

Now, we can give our new geometry a bend. In *Figure 3.29*, we can see how it should look.

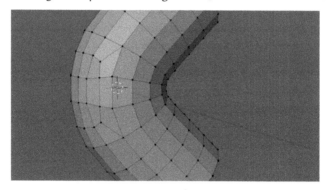

Figure 3.29 – Bending our new geometry

Now that we can see it deformed, it looks pretty good. The quads are all mostly the same shape, and they are not being warped too drastically – that actually makes a lot of sense. If you recall our last example with the plane, the edges of the plane were much bigger than the middle ones, and the faces that transferred from the small ones to the big ones were small on one side and big on the other. If we compare a gradient of faces between a plane and a cylinder with the same topology, it is a bit easier to see. You can see a comparison in *Figure 3.30*.

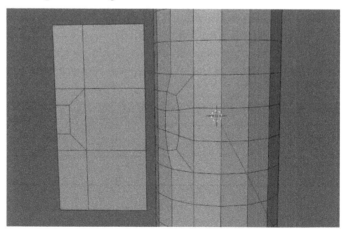

Figure 3.30 – Comparison of a plane (left) and a cylinder (right) with the same topology

The small faces are on the back so that they can be stretched out. The transitioning areas are on the sides, where they will transfer between the two sizes. The big faces are on the inside of the joint, where the faces are going to compress.

With that, we have managed to solve our deformation issues with the simple elbow joint. Now that we know how to deform our mesh with simple stretching and compressing, we can focus on some more joints.

Applying the twisting deformation rule

Twisting deformations are usually used on any shape that has one side rotate around an axis, while another part of that mesh either does not rotate or rotates the other way. You can see the result of this twisting happening to a cylinder in *Figure 3.31*.

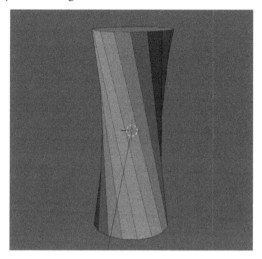

Figure 3.31 – Twisting deformation on a cylinder

Notice how the edges connecting the top and bottom vertices are slanted. To achieve this, select the top face of the cylinder and press *R* and *Z* to rotate only the top face. Notice how we rotated them along an axis in line with the vertical edges, and perpendicular to the edges going around the cylinder. That is our **second deformation rule**, that the axis of twisting should be in line with the edge flow.

We can also use our good old grid to test this out too. In *Figure 3.32*, we can see our plane properly deformed along the edges in line with the red *x*-axis.

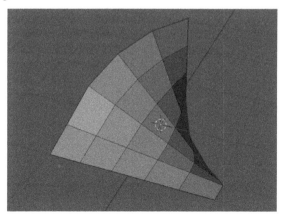

Figure 3.32 – A plane twisted around the x-axis

You will notice on both the cylinder and the plane that there are limitations to how far you can twist certain shapes. This is usually caused by too little geometry to work with. You can see the issue in *Figure 3.33*.

Figure 3.33 – A twisted cylinder with insufficient geometry

The preceding figure is reminiscent of the twisted end of a candy wrapper, which is why this problem is referred to as **candy wrappering**. It can be helped by adding more geometry to distribute the twisting across the faces a little bit more smoothly. If we look at *Figure 3.34*, we can see how this might look on a cylinder.

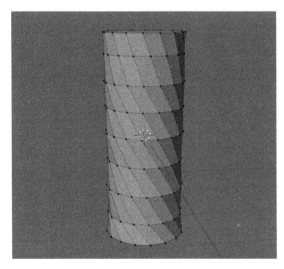

Figure 3.34 – A twisted cylinder with geometry added to smoothen faces

You might be able to notice the top and bottom loops of vertices are in the same place as in *Figure 3.33*. The only difference is in the middle, where we relaxed some of the loops by rotating the faces in the other direction.

Applying the twisting rule to shoulder deformations

Cylinders represent some of the easier shapes you can twist. It is much more difficult to have a joint twist within a hinged joint. These sorts of twists usually happen when a connection between two grids occurs. These sorts of connections are most commonly used to connect limbs to a body – for instance, the hip joint or the shoulder joint. Even the neck has a joint like this. *Figure 3.35* shows us a simple shoulder topology.

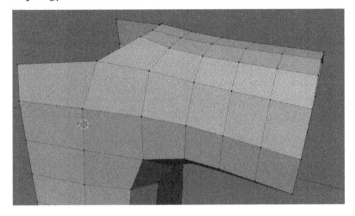

Figure 3.35 – A shoulder topology

First off, we should look at our topology to see what it is doing. In this example, the front topology matches the back topology. At a first glance when looking at where the arm meets the shoulder, we can see a pole right where the armpit starts highlighted in *Figure 3.36*.

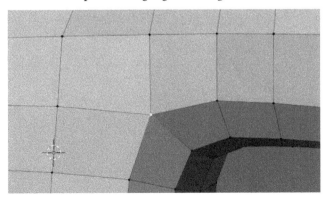

Figure 3.36 – A pole where the armpit starts

Because our mesh is mirrored front to back, and because we have one on the front, that means that there is another one on the back. If you remember, a pole indicates a corner of the intersection of two girds.

That means that we have two poles connecting the two grids. If we were to visualize this topology as straight grids, it would look something like *Figure 3.37*.

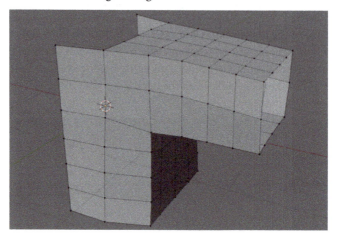

Figure 3.37 – Visualizing topology as straight grids

To improve the deformations around the pole where the arm meets, we are going to move the pole further away from the area of deformation. You can see the pole moved away from the joint in *Figure 3.38*.

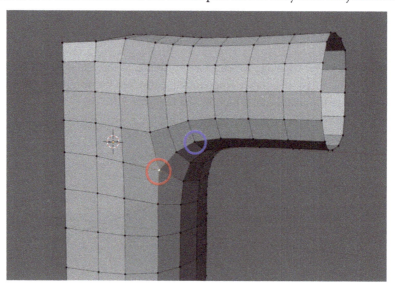

Figure 3.38 – Result of moving the pole further away from the joint

The blue circle marks where the pole used to be, and the red circle highlights the new position of the pole. Why exactly we do this may not be obvious at first, but with a few examples, it will hopefully make a bit more sense. In fact, this leads straight into our next rule.

Applying the intersecting grids deformation rule

Moving this pole is an example of the **third deformation rule**. Always have a buffer of at least one row of faces between a pole and the specific areas of deformation. You can see our new topology deforming in *Figure 3.39*.

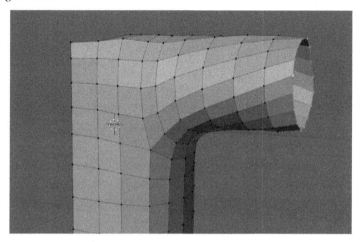

Figure 3.39 – Our new deformed topology with the third deformation rule applied

It is immediately noticeable that the transition is much smoother now that we have moved that pole further away from the area of deformation. This gave the geometry more room to do the deformation and reduced the amount of deformation the pole was experiencing as a result. Because a pole has edges pointing in more than just four directions, it is difficult to get it to deform cleanly. In *Figure 3.40*, you can see a plane with a regular four-edged vertex and another plane with a five-edged pole being deformed side by side.

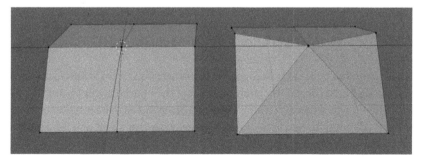

Figure 3.40 – Comparison between a regular four-edged point (left) and a five-edged pole (right)

Because none of the edges are lined up with each other on the pole, it does not deform nicely like a normal quad.

You might ask yourself why you cannot just line up two of the edges so that they can deform cleanly. To a certain extent, you can do this, as *Figure 3.41* shows us.

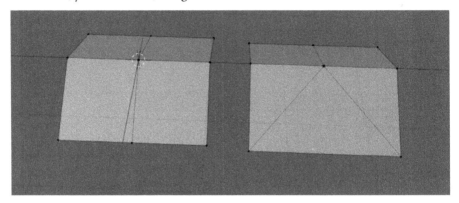

Figure 3.41 – Lining up edges to improve deformation

While aligning the edges in this fashion will work in this specific situation, as soon as we want to deform the mesh in any other way, this trick falls apart. This is shown in *Figure 3.42*.

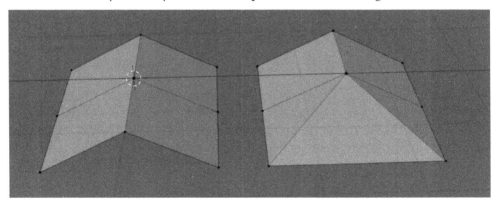

Figure 3.42 – Deforming the mesh further makes lining up edges an inadequate solution

Instead of deforming around our lined-up edges, we deformed along a perpendicular axis. While we have one edge pointing in the correct direction on the top, the bottom edges form a V, causing it to behave erratically. That is why when deforming a pole, we always need to be mindful of the directions of deformation that are going to be applied to it. That way, we can avoid any unwanted topology artifacts caused by the mesh bending unpredictably.

Applying the intersecting grids rule to hip deformations

Now, we can try to apply this same way of thinking to the hip. You may remember making this shape in the final section of *Chapter 2*. Now that we have learned a few more things about how we approach deformations, we should take another look at that topology again to explain some of the reasons we made it the way we did. To start, we are going to look at another way we might approach the topology of that shape in *Figure 3.43*.

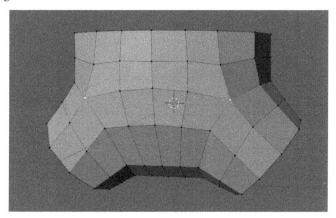

Figure 3.43 – The old example of the topology of this shape

This topology is identical to the topology we covered in the previous chapter, except for one difference. The pole in this example is positioned directly on the area of deformation. As we just learned from our previous rule, there needs to be a row of faces between a pole and the area of deformation. If we move our pole away from this area, you can see the solution to this in *Figure 3.44*.

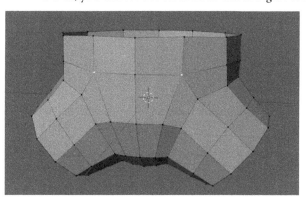

Figure 3.44 – A new perspective of the hip topology with the pole positioned on the area of deformation

Now that we have applied the three rules of deformation to common shapes other than a plane, let us see how they relate to a simple cloth simulation.

Topology for cloth simulations

Just like normal deformations, cloth and soft body simulations follow the three deformation rules as well. While we will not get into the specific settings to apply to a soft body mesh, it is important to understand how the topology is going to react.

To start, subdivide a plane to make sure that our mesh has enough geometry to work with. Because the cloth simulation can be affected by the modifier stack, we should do as much of the subdividing as we can in the **Modifiers** tab. Be careful with the **Subdivision** modifier though – adding too many subdivisions can massively slow down your computer or outright crash Blender. We can see the **Subdivision** modifier in *Figure 3.45*.

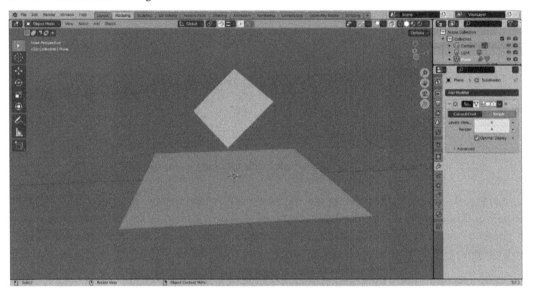

Figure 3.45 – Subdividing the plane

To start out, do the following:

1. Add a **Subdivision** modifier set to **Simple**. This allows us to add more geometry without also smoothing the mesh.

2. To add a **Cloth** modifier, open the **Modifier** dropdown in the **Modifiers** tab.

3. Look under the **Physics** section.

4. Select the **Cloth** modifier. You can see this tab in *Figure 3.46*:

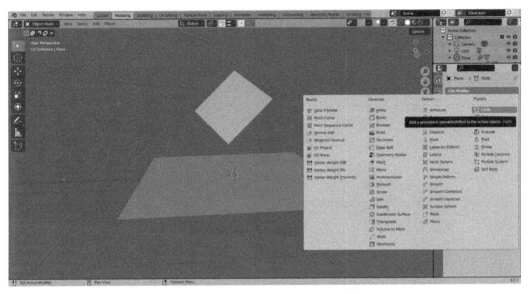

Figure 3.46 – Selecting the Cloth modifier

This will give us a simple cloth simulation when we run the simulation. The biggest determinant of the quality of a cloth deformation is the number of faces it has to work with. In *Figure 3.47*, you can see the same cloth settings run on meshes with different subdivision levels.

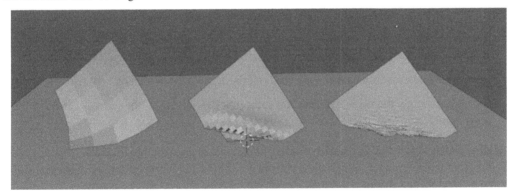

Figure 3.47 – Using the cloth setting on meshes with different subdivision levels

The left-hand model has two levels of subdivision, the center model has four, and the right-hand model has six. Of course, the more subdivisions you have, the longer the simulations will take – so what can we do to get the deformations we want without hiking up the simulation times?

Without touching the cloth settings, there are a few things we can do. The easiest option is to add another subdivision after the cloth modifier. In *Figure 3.48*, you can see the meshes with two subdivisions applied.

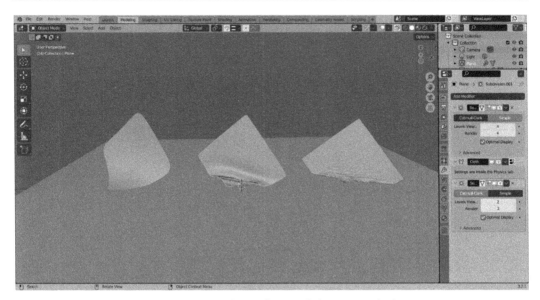

Figure 3.48 – Meshes with two subdivisions applied

Thus, one of the benefits of the cloth simulation being affected by the modifiers is that we can make changes to the simulation after the simulation is calculated. This means we can add as much detail as we want without having to worry too much about how it impacts the simulation speed. Ultimately, there are three things to worry about with simulations:

- You need to have really evenly sized quads to make sure that you get a consistent simulation. That means quads with edges that are even in size, and as close to square as possible.

- You need to make sure you have enough quads for the detail of the simulation you are trying to do.

- You need to apply the three deformation rules to it as well, especially for simulations with fewer quads.

This is all we will be discussing in relation to simulations, but it is important to touch on these to have a full understanding of topology, especially if you are going to be modeling clothes on characters and therefore using cloth simulations.

Summary

In this chapter, we learned how to approach a deformation problem. We also learned about the three deformation rules, including why the edges of our grid need to be parallel to the axis of deformation, why the axis of twisting should be in line with the edge flow, and why we must always ensure a buffer of at least one row of faces between a pole, and the specific areas of deformation. We also learned how to prepare a mesh for cloth simulation deformations.

Now that we know what rules to follow when deforming our topology, we can focus on our last major consideration before moving on to some real applications. In the next chapter, we will be looking at how UVs are affected by our topology, and how that can show us issues we may not have seen before.

4

Improving Topology Using UV Maps

With the basics of topology and how it deforms taken care of, there is one last consideration to be made when laying out the topology of a model. That last consideration is how the material is going to be applied to it. There are a few ways to do this, but the most consistent way of doing this is by UV unwrapping and applying image textures to the surface of the model.

This chapter will start by introducing you to a UV map, and then we will move on to explaining how it works. Once we have an understanding of the UV map itself, we will move on to its relationship to the topology of the model. Then, we will learn how to check a model's UV map and improve it.

In this chapter, we will be learning about the following subjects:

- What is a UV map?
- Ensuring clean topology for clean UV maps
- Unwrapping the mesh
- Improving unwraps

What is a UV map?

To start out, what is a UV map? A **UV map** is a 2D representation of a 3D mesh. It is where you take all of the faces of your mesh and flatten them out. In *Figure 4.1*, you can see a cube next to its UV map:

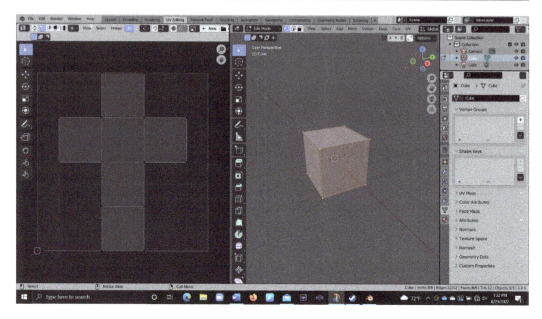

Figure 4.1 – UV map of a 3D cube

To view your UV map, it is easiest to navigate to a different workspace at the top of your screen. The workspace you are looking for is the **UV Editing** workspace. You can see this workspace in *Figure 4.2*:

Figure 4.2 – UV Editing workspace

Now that we are in the UV Editing workspace, your screen should match the layout of *Figure 4.1*. To see your UV map on the left, go through the following steps:

1. Select the object in **Object Mode**.

2. Then switch to **Edit Mode**.

3. And finally, press *A* to select all of the faces of the model.

By default, you can only edit the UVs of faces that you have selected.

With all of the faces selected, you should now see your UV map on the left. You can now select any of the edges, vertices, or faces on your UV map to modify it. All of the same transform hotkeys that you use in the viewport also apply here: *G* to move, *S* to scale, and *R* to rotate. *Figure 4.3* shows a vertex on the UV being moved:

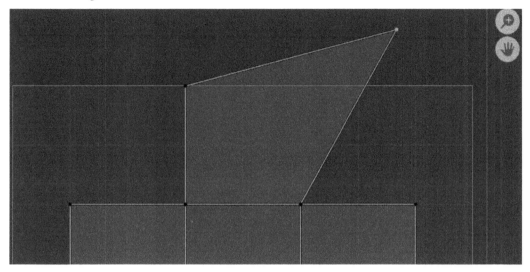

Figure 4.3 – Editing the UV faces

You may have also noticed that navigation in this area is also similar to the viewport. Use the *MMB* to pan and scroll up and down on the mouse wheel to zoom in and out.

At the moment, it is difficult to tell how our model is being affected by modifying our UV map, so we are going to add a simple material. That way, we will easily be able to see how our UV relates to our mesh. First, we will press the tab labeled **Image** in the **UV Editing** area, as shown in *Figure 4.4*:

Figure 4.4 – Image tab in UV editor

In that dropdown, select **New**. After you select **New**, a tab will open showing you the parameters of the image you want to create. The part we want to focus on is the **Generated Type** option. This allows us to choose from some preset images called textures that we can use for UV mapping. The selection we want to choose is the **Color Grid** option from the dropdown shown in *Figure 4.5*:

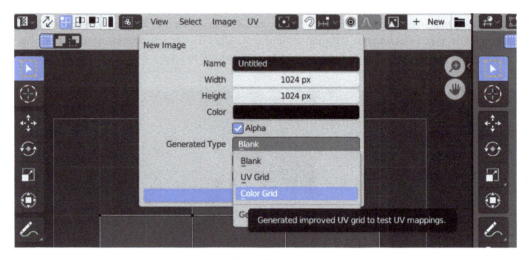

Figure 4.5 – Choosing the Color Grid option for the new image texture

With this setting change, we can click on **OK** at the bottom of the **New Image** tab and finalize our image texture. At the moment, we have only created the image texture in Blender, and we need to save it to our computer. To save the image, navigate back to the same **Image** tab we pressed to create the texture. Now that we actually have an image in the UV editor, we have a few more options. In the **Image** dropdown, select the **Save As…** option, as shown in *Figure 4.6*:

Figure 4.6 – The Save As… option

With the image made and saved, we need to apply it to the model by assigning it to the material of the model in the Shader Editor.

Applying a texture

To easily switch to the proper editor, we will use another workspace. To modify materials, we are going to use the **Shading** workspace shown in *Figure 4.7*:

Figure 4.7 – The Shading workspace

At the bottom of the Shading workspace, we have the main editor type called the Shader Editor. The **Shader Editor** allows you to modify the materials of an object. A **Material** is a set of parameters that determine how your model interacts with light. The individual processes are contained in nodes, and the main nodes that determine how you want your material to look are called **shaders**. In *Figure 4.8*, you can see two nodes next to each other in the Shader Editor:

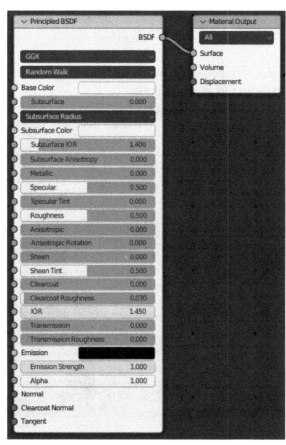

Figure 4.8 – The Shader Editor

The node on the left is the shader node, called **Principled BSDF**, and on the right, we have **Material Output**. **Principled BSDF** is a shader node with the general settings you would have to set for most materials inspired by a similar workflow used by *Pixar*.

The **Material Output** node determines what is actually being displayed on the surface of the model. If your shader node is not plugged into **Material Output**, it will not appear on the model. With that short description out of the way, we can add our image to the material. Before we can actually add it to the material, we need to bring it into the Shader Editor. To do this, follow these steps:

1. Start by opening the **Add** tab at the top of the Shader Editor, as shown in *Figure 4.9*:

Figure 4.9 – Add option

2. In this dropdown, hover over the **Texture** option, and then click on the **Image Texture** option. *Figure 4.10* shows you what this should look like:

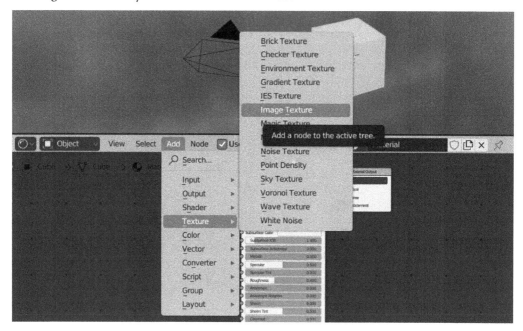

Figure 4.10 – Image Texture option

3. As soon as we select **Image Texture** from the menu, a new **Image Texture** node will appear in our Shader Editor. You'll notice that the node is following your mouse. To finalize its position, press *LMB* when you are satisfied with its placement. To move it again, just like in our viewport and UV editor, select the image texture node with *LMB* and press *G*.

4. With our texture node in the Shader Editor, we need to assign our texture to it. To do this, simply click on the image icon on our **Image Texture** node, as shown in *Figure 4.11*:

Figure 4.11 – Assigning a texture to the node

5. Finally, all that is left to do is plug our texture into the shader. To connect our nodes together, press and hold *LMB* on the yellow circle on the right side of our texture node, as shown in *Figure 4.12*:

Figure 4.12 – Yellow color output

6. When we move our mouse around now with *LMB* held down, we get a line, as shown in *Figure 4.13*:

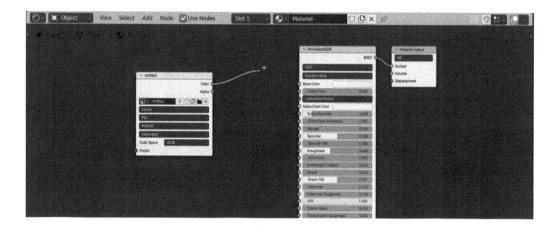

Figure 4.13 – Plugging the texture

All we need to do now is drag that line over to the **Base Color** input on the **Principled BSDF** shader node, and release *LMB*. Now, you should have a line going from the **Color** output of the **Image Texture** node, going into the **Base Color** input of the **Principled BSDF** node. We can see this connection in *Figure 4.14*:

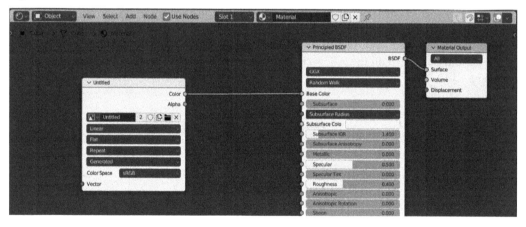

Figure 4.14 – Successful connection of the Color node to the Principle BSDF node

With our texture finally connected to our material, we should be able to see it on our model in our viewport.

Shading modes

If your cube and background still look gray, that is because you need to change your Viewport shading mode. The **Viewport shading mode** determines how your models appear in the viewport. You can see the four shading mode selections on the top right of the viewport in *Figure 4.15*:

Figure 4.15 – Viewport shading modes

The leftmost option is the Wireframe shading mode. The **Wireframe shading mode** shows an outline of all of the mesh edges, but not the faces, allowing you to see through the mesh. If we look at our cube in Wireframe shading mode, we will see an effect similar to that in *Figure 4.16*:

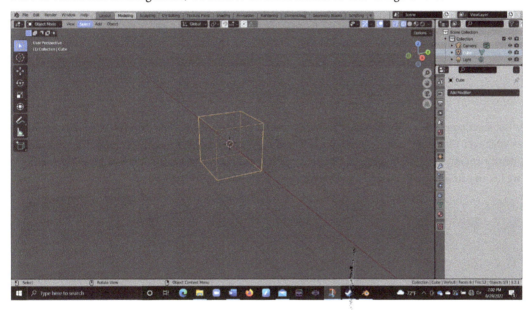

Figure 4.16 – Wireframe shading mode

The next one to the right of the Wireframe shading mode is the Solid shading mode. The **Solid shading mode** is the shading mode we have been using up until this point. It gives us a solid color with basic shading to view the contours of the model.

The next shading mode to the right of the Solid shading mode is the Material Preview shading mode. The **Material Preview shading mode** uses the **EEVEE** rendering engine to add a quick lighting effect to preview materials, just as the name suggests. By default, it switches to Material Preview when you enter the shading workspace. You can see the Material Preview shading mode in *Figure 4.17*:

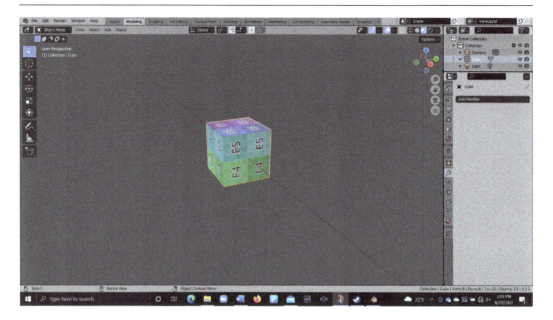

Figure 4.17 – Material Preview shading mode

The last shading mode on the far right is the Rendered shading mode. The **Rendered shading mode** uses whatever rendering engine you have selected to render the viewport. By default, the real-time rendering engine, **EEVEE**, is selected, but you can change this to **Cycles**, Blender's ray-tracing engine, as well. You can see the Rendered shading mode in action in *Figure 4.18*:

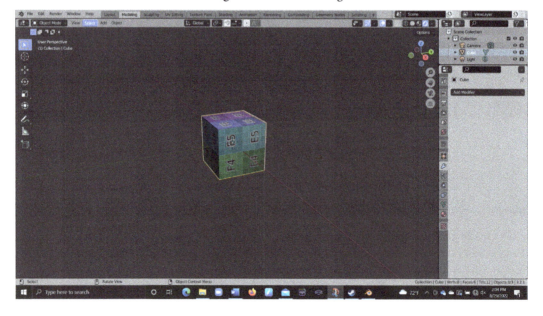

Figure 4.18 – Rendered shading mode

Now, all we need to do is see how our texture is affected by changing our UV map; we will go back to **Material Preview**.

To quickly switch between the shading modes, press and hold Z until the pie menu in *Figure 4.19* pops up:

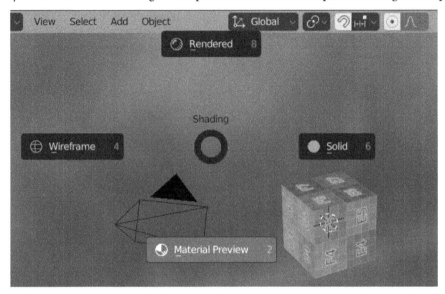

Figure 4.19 – Switching between shading modes

We can move our mouse around to select one of these, and once we have our desired view selected, all we have to do is release the Z key.

Manipulating the UV map

With our texture applied, and the ability to view it, we can start messing with our UV map. First, we are going to go back into the **UV Editing** workspace at the top of our screen. Now, we can hold down Z to select our **Material Preview** shading mode. Next, we want to press these opposing arrows at the top left of our UV editor area shown in *Figure 4.19*, called **UV Sync Selection**:

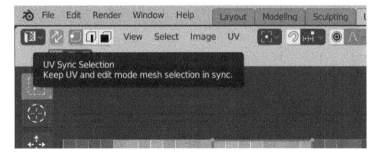

Figure 4.20 – UV Sync Selection

Before enabling this setting, we needed the entire mesh selected in Edit Mode to see the whole UV map, but now we can see the map without selecting the mesh, though we still need to be in Edit Mode.

This also syncs our selections between our UV editor and our viewport. Now, If I select a vertex on the UV editor, its corresponding vertex on the mesh is selected. You can see this selection in *Figure 4.21*:

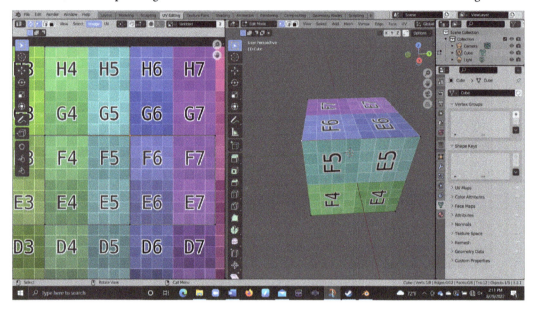

Figure 4.21 – Synced selections

You will also notice that if you select specific individual vertices in the viewport, multiple vertices are selected in the UV editor. That is because these vertices fall on edges that are split in order to flatten out the mesh, so the selected vertices in the UV editor share a vertex in the 3D Viewport. This is illustrated in *Figure 4.22*.

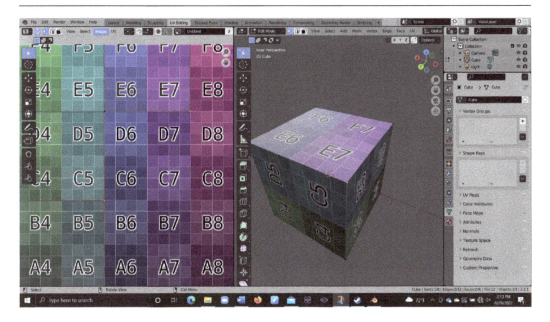

Figure 4.22 – One selected vertex in the 3D Viewport selecting two in the UV Editor

All of the vertices on the UV map represent the same vertex on the mesh. By selecting the vertex in the viewport, it will select all of the corresponding vertices in the UV editor. To avoid this behavior, we just have to deselect the icon in the top left with the opposing arrows shown in *Figure 4.23* and select the whole mesh again.

Figure 4.23 – UV Sync Selection icon

If we move one of the vertices in the UV editor, you can see that it changes the way the texture is laid out on our cube. You can see this in *Figure 4.24*:

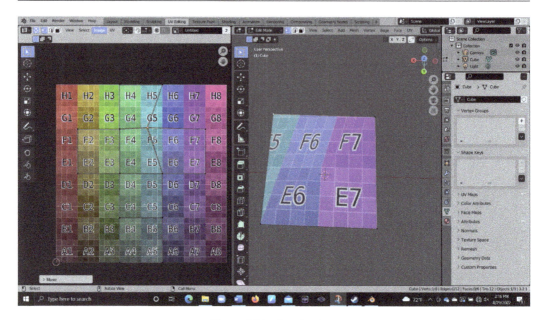

Figure 4.24 – Deselection

The faces in the UV editor have a one-to-one relationship with the faces on the mesh, so if we warp our faces in the editor, it will warp how the texture is displayed on the mesh. You can also select the whole UV map and move it around to see how the two are related.

Now, we know how to add an image texture, how to apply that texture to a model using the Shader Editor, how to see that texture on the model by changing the shading mode, how to apply the UV Sync option, and understand how moving vertices in the UV map changes the way the texture is displayed on the model. With this understanding of the basics of a UV map, we can start looking at how we should set up our topology to make manipulating this UV map as easy as possible.

Ensuring clean topology for clean UV maps

Thankfully, there is not too much more we need to think about when laying out our topology for UV maps so long as we have followed our other topology rules. So first, we will look at the rules that apply to UV mapping most. One of the more painful topology issues to work around is when the topology is not flowing with the geometry properly. In *Figure 4.25*, I have two planes that I want to unwrap:

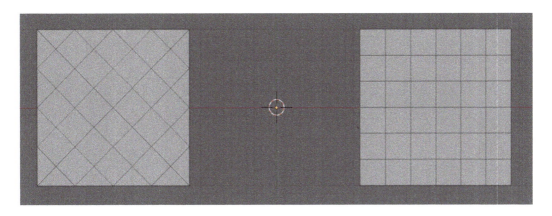

Figure 4.25 – Planes for unwrapping

The one on the left has jagged edges going left to right, and the plane on the right has straight edges like a normal grid. If we split these in half and look at them in the UV editor in *Figure 4.26*, you can see how the jagged edges intersect our texture at awkward angles:

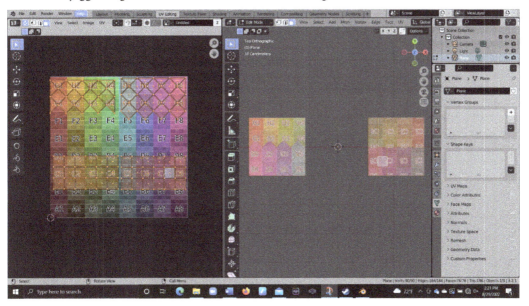

Figure 4.26 – Jagged UV map (left), straight UV map (right)

If we wanted the grid pattern to line up perfectly with our plane, we would not be able to use a jagged pattern like the plane. This is specifically an issue if you are setting up your textures to work with a tiling texture, where you need your UV map to unwrap into a perfect plane.

As the name might suggest, a **tileable texture** is an image texture that can be laid next to itself like a tile floor without having any visible seams between its tiles. You can see an example of a tileable texture in *Figure 4.27*:

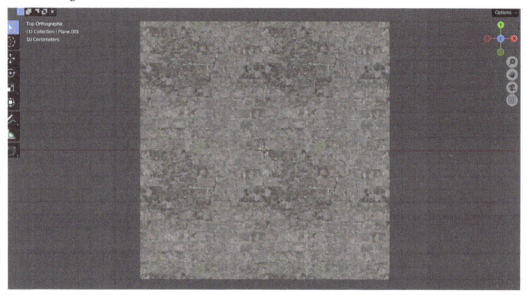

Figure 4.27 – Tileable texture

While you may be able to see the texture repeating, the edges all blend into themselves. In *Figure 4.28*, you can see the texture applied to our planes:

Figure 4.28 – Applied texture on the planes

The plane with the straight UV map can be manipulated to tile seamlessly, but the jagged one cannot.

When using tileable textures, we want to try to avoid these messed-up edges as much as possible. Unfortunately, it is not always avoidable. In these instances, you want to try to minimize the effect it has on the final model. Like the topology rule that says that the topology should follow the geometry, the rule that says that loops need to terminate into themselves is just as important.

Both of these rules are very easy to spot when UV unwrapping because they will affect the textures immediately. In *Figure 4.29*, you can see a cylinder that is connected at an offset, so the faces and UV map do not line up with each other:

Figure 4.29 – Misaligned unwrapped texture

If we look at our mesh and UV map, it is easy to tell that the texture is not lining up with itself and that the texture actually spirals around the mesh as well. We could try to make the texture line up in the UV editor, as in *Figure 4.30*:

Figure 4.30 – Lining up the texture

This way, we would even be able to tile the texture without any seams. Unfortunately, this warps the texture because we are forced to stretch our UV map to match up the edges. When a face mismatch like this takes place in a more complex model, it is unlikely that we would even be able to get it to line up. When looking at simple shapes like this, it is easy to understand how and where to split the mesh when flattening it out, but real shapes often have more complex shapes that you need to flatten out. To do this, we need to lay out our topology for unwrapping, which we will see in the next section.

Unwrapping the mesh

Now that we understand a little bit about what UV maps are, and why getting them right is important, we can learn how to actually unwrap a mesh. To unwrap a mesh, we start by marking the edges we want to split our mesh along. To do this, follow these steps:

1. Start by selecting the edge we want to split. Next, press *Ctrl + E* to open up the **Edge** menu and select **Mark Seam** from the dropdown, as shown in *Figure 4.31*:

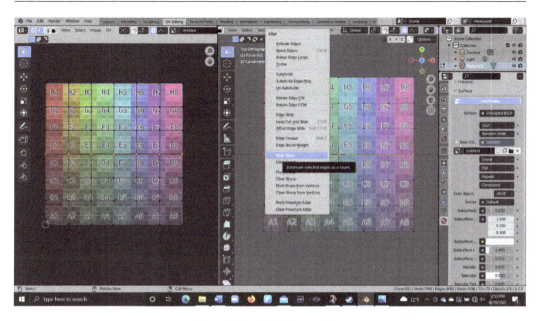

Figure 4.31 – Mark Seam option

After marking the seam, a red line should appear along the selected edge, as in *Figure 4.32*:

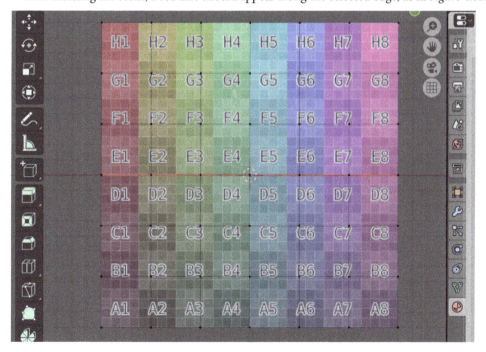

Figure 4.32 – Marked seam for unwrapping

2.   This seam shows where our model will split when unwrapping. To unwrap, we start by selecting our whole mesh with *A*. With everything selected, we can press *U*. This brings up the **UV Mapping** tab, as shown in *Figure 4.33*:

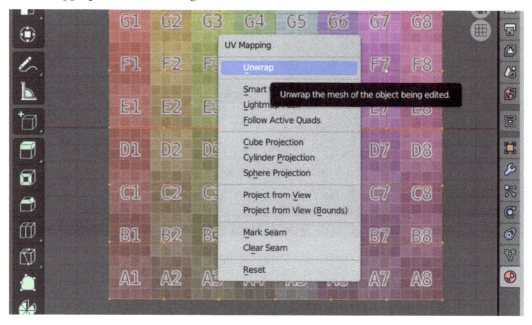

Figure 4.33 – UV Mapping tab

This is the menu that allows us to decide how we want to unwrap our mesh. When manually marking and unwrapping our mesh, just pressing **Unwrap** at the top of the dropdown is what we want to do. After we unwrap our mesh, we can see it in the UV editor and viewport in *Figure 4.34*:

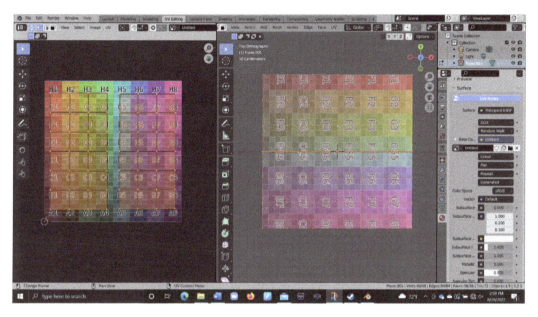

Figure 4.34 – Unwrapped mesh

You will notice in our viewport that we still have our red line indicating where we marked our mesh, and in the UV editor, we have two sections indicating the split. Next, we are going to look at a more complex mesh and see how we would approach unwrapping that. A good rule of thumb to start with is to keep our seams on corners, creases, or the intersections of grids.

Our two major objectives when unwrapping are to hide seams and reduce warping when unwrapping to 2D. You will likely recognize the model in *Figure 4.35*, as we have used this shape a few times now:

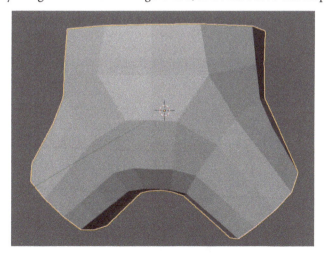

Figure 4.35 – A complex model being used for unwrapping

This is going to be the model we will try to unwrap for the first time. To start, we are going to look at that first rule and unwrap where the two grids intersect. If we look at *Figure 4.36*, we can see where the grid from the T shape and the cylinder of the leg intersect:

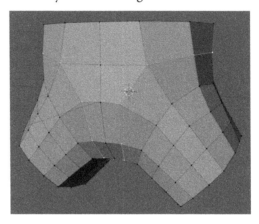

Figure 4.36 – Selected edges at the intersection of the leg and hip

While this is where the two grids intersect, we are not actually going to put our seam right on this intersection. That is because there is not a good crease or corner at that intersection to actually put our seam to try and blend it in. So instead, we are going to move our seam to the actual joint of the two shapes. As a recap, to mark the seam, we should go through the following steps:

1. Select our seams in **Edit Mode**.
2. Press *Ctrl + E* to open up the **Edge** menu.
3. Navigate down to the **Mark Seam** option.
4. After selecting that, the **Edge** menu should close and you should have seams as shown in *Figure 4.37*:

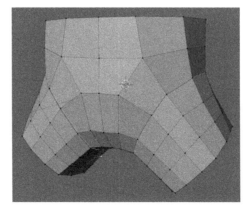

Figure 4.37 – Seams added to the model

With our seams in place, we can now try to unwrap our mesh for the first time. To UV unwrap the mesh, follow these steps:

1. Select the whole mesh in **Edit Mode**.

2. Press *U* to open up the **Unwrap** menu.

3. Select **Unwrap** from the top of the list.

With our mesh unwrapped, it should look like *Figure 4.38*:

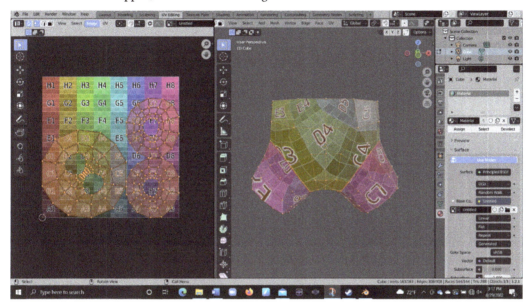

Figure 4.38 – Unwrapped mesh for our model

You will notice that the unwrap is severely warping our mesh even though we unwrapped around the intersections of the grids. That is because after separating our grids, we still need to unwrap those individual grids so that they will unwrap as close to a normal flat grid as possible.

To start with, we can use the UV editor to help us mark seams. To do this, make sure you have the **UV Sync Selection** option enabled, as in *Figure 4.39*:

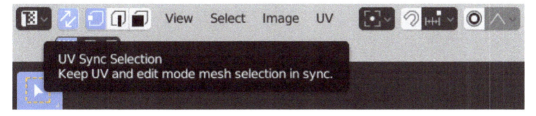

Figure 4.39 – Enabling the UV Sync Selection option

This is the option that syncs selections between the viewport and UV editor. This makes it a bit easier to see what edges we are selecting to get marked. In the **UV Editing** tab, select one of the loops along each of the legs that we separated with our last seams. Make sure that these seams are matching on either side. We are also going to try and hide them as much as possible, so we can tuck them on the inside so that they are harder to see. *Figure 4.40* shows what this selection looks like in the viewport and the UV editor.

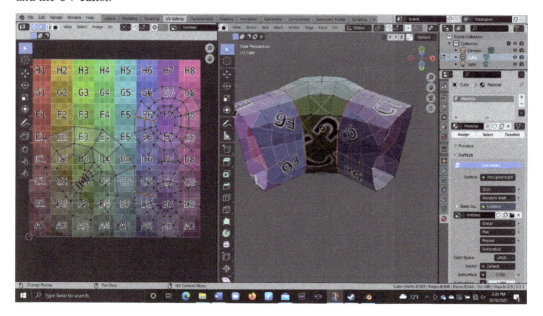

Figure 4.40 – Selected seams in the UV editor

With our edges selected, we can press *Ctrl + E* and mark our seams and then unwrap with *U*. Now, our legs should be nicely unwrapped into normal grids, and our pattern hardly looks warped on the legs. Next, we need to do the T section in the middle. To get this to play flat, we are going to have to mark at least two of the three connections forming the shape. In *Figure 4.41*, you can see one of these selections.

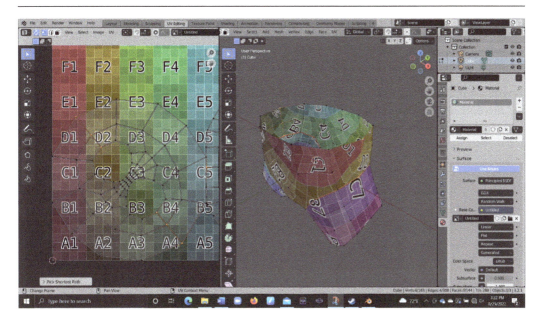

Figure 4.41 – Marking two connections

Again, we are going to mark our seams with *Ctrl + E*, then unwrap the mesh with *U*. With our mesh unwrapped using these new seams, we can see the final unwrap in *Figure 4.42*.

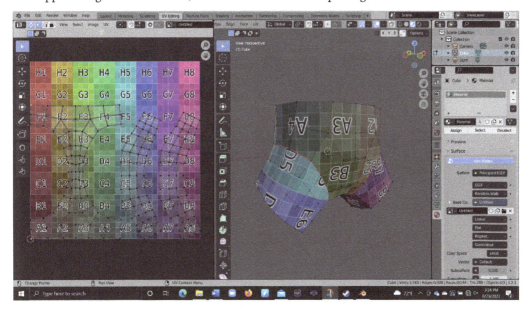

Figure 4.42 – Final unwrap after using new seams

Now, when we look at our mesh in **Object Mode** with a texture applied, nothing is terribly warped, and all of our seams are as hidden as they can be if we are going to unwrap like this. Another method to unwrap this shape is along the sewn seams that clothing would have.

This requires our topology to have edges that also follow the clothing seams. For a normal pair of pants, there are usually four major seams that hold the pants together. These seams run up the inside, outside, and crotch of the pants. You can see these seams marked in *Figure 4.43*.

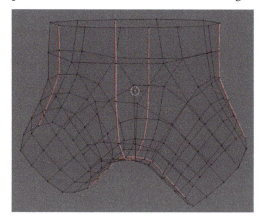

Figure 4.43 – Seam mapping

When we unwrap the mesh, it should look something like *Figure 4.44*:

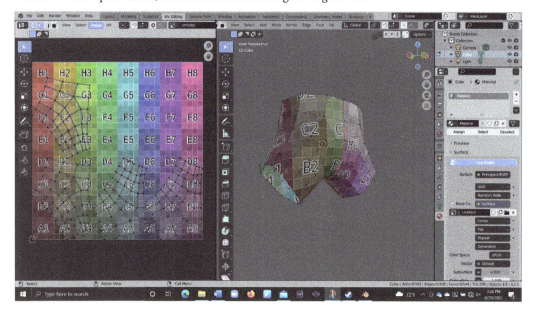

Figure 4.44 – Unwrapped model

This mesh also has minimal warping to the texture on the mesh and displays the grid nicely on the mesh. We can take a closer look at the quality of our unwraps by introducing a new tool in the next section.

Improving unwraps

Our new tool is located in the **Overlays** tab of the **UV Editing** area. You can see the **Overlays** tab open in *Figure 4.45*:

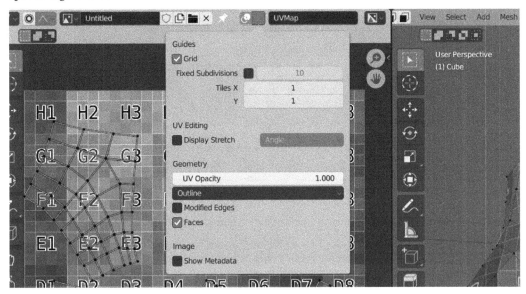

Figure 4.45 – The Overlay tab

In this tab, we can scroll down to the **Display Stretch** checkbox shown in *Figure 4.46*:

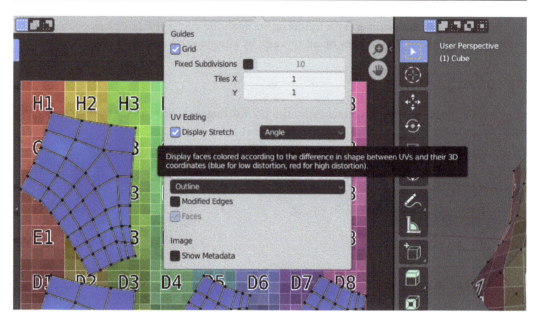

Figure 4.46 – The Display Stretch option

With this setting, we can check two things: the area of the individual faces, and the amount of deformation each of the faces has. By default, it checks for the angle, but this can be changed in the **Overlays** tab we just used to enable the setting. With this setting set to **Angle**, we can see how it affects a cube in *Figure 4.47*.

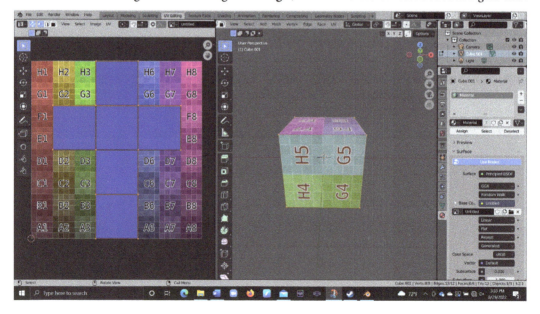

Figure 4.47 – Affected cube model with new settings

Notice how all of the faces are blue. This means that there is no warping because all of the faces are being warped from how they are actually shaped on the mesh. Notice how if we move a point on the mesh, or if we move a vertex in the UV editor, the UV map starts to turn green, and then red on the vertices it is affecting. *Figure 4.48* shows a vertex in the UV editor being moved.

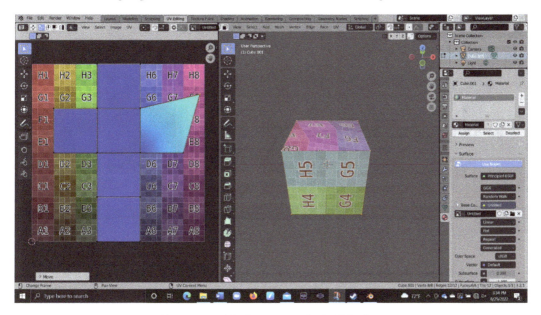

Figure 4.48 – Moving a vertex in the UV editor

This can be corrected by either moving the mesh or the UV map to match the shapes as much as possible. When unwrapping a finished mesh, that usually means tweaking the UV map. This is often a good indication of whether you have enough seams on your UV map, or whether you have them placed well enough. If you unwrap your mesh and there is a dramatic amount of warping, you will likely have to change your unwrap.

The final thing you want to think about when unwrapping your model is something called texel density. **Texel density** is the size of the face in relation to how much space it takes up on the texture. If we look back at our pants example, with the separate UV islands, we can see this in practice. In *Figure 4.49*, we can see the two legs next to each other with different texel densities.

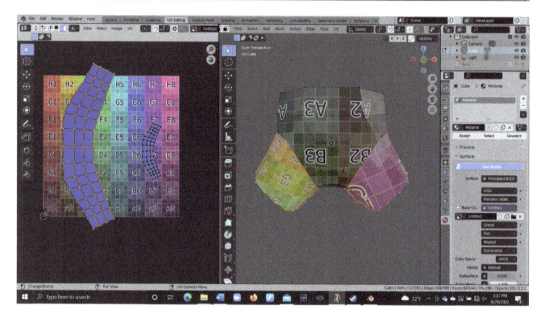

Figure 4.49 – Different texel densities

The leg on the left has a shrunken UV. This reduces the texel density of that leg because it is taking up less space on the texture and, by extension, makes the texture look bigger on the mesh. The leg on the right has an enlarged UV. This increases the texel density and makes the texture look smaller because it takes up more space. This is how you can control what parts of the mesh get more detail than others. Have a higher texel density on the parts you want more detail on, to allow them to take up more space on the texture. This is something we need to control on each of our unwraps, and we need our mesh to have as consistently sized quads as possible to help with this.

Thankfully, if we observe all of our previous topology rules up until this point, our only trouble should be in marking our seams properly, but if you find that you are unable to get the result you want regardless of the seams, it likely means that you violated a rule earlier on.

Summary

In this chapter, you learned how to mark seams, UV-unwrap your mesh, and apply a texture to a mesh to check your UV. We also understood where to put our seams when unwrapping and which topology rules affect UV unwrapping the most. By now, you should be able to use UV unwrapping to check your topology and know how to check the quality of your UV map.

UV unwrapping is the last point to check your topology before doing all of the materials and rigging if your model calls for those, so it is important that you catch any major issues here. Otherwise, you may need to redo a tremendous amount of work. With these last considerations you need when preparing for UV unwrapping taken care of, we should be ready to go straight into a more complex model. In the next chapter, we will take a look at a character's head, and apply all of the topology rules we have learned in this chapter to that model.

Part 2 –
Using Topology to Create Appropriate Models

With a foundation in topology rules established in the first part, this section is dedicated to putting the rules into practice. We will apply our rules to the two types of models that you may approach when using topology, organic and hard surface models.

We will start with a humanoid character to practice our topology on an organic shape. The character will be broken down into segments to be approached individually. Once we have been through the topology process on the organic models, we will repeat the process on a blaster acting as our hard surface model. Then, we will finish with how to optimize the finished models.

By the end of these chapters, you will know how to separate models into sections that you can focus on individually. We will have gone through examples of both hard surface and organic models. Then, we will learn when, and how, we can bend topology rules to optimize our models.

This part of the book comprises the following chapters:

5

Topology on a Humanoid Head

Now that we have all of our topology constraints out of the way, we can finally try to apply them to a more complex model. This next part of the book will be focusing on just that, applying the rules that we learned in the previous four chapters and using them to create the topology for different models.

In this chapter, we will be introduced to the retopology of a humanoid character. We will start off with a brief description of retopology. We will start our topology with the head. This chapter will establish a process when laying out our topology that we will apply to all of our models in the future. We will focus on the areas of detail first, then work on connecting them together. In the case of the head, we will start with the face. Then on the face, we will break it down into its separate features.

In this chapter, we will be covering the following subjects:

- Introduction to retopology
- Retopology of the face
- Retopology of the ear
- Retopology of the back of the head

Introduction to retopology

To start off, we should talk about retopology. **Retopology** is a process in which you modify or completely redo the topology of a mesh. In *Figure 5.1*, you can see a mesh with bad topology before retopology on the left, and then after retopology on the right:

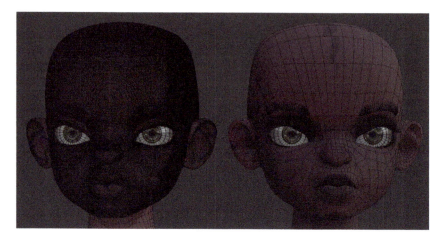

Figure 5.1 – Comparison of mesh with bad topology (left) and mesh after retopology (right)

When approaching a retopology, it can be helpful to look at good examples of the process. You can see one of these references in *Figure 5.2*.

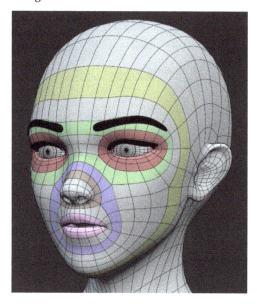

Figure 5.2 – An example of a topology reference that I have made

These are nice to get an idea of where our defining loops might be, but when modeling anything other than a basic human, the topology will likely be different. That is why we chose a bit more of a unique model to retopologize. This will give us a chance to look at the reasons we put the topology where we do and form a practical understanding through the application of our topology rules.

Revising topology checks and rules

It might be helpful if we list these rules all in one place, as we have learned quite a bit in the previous four chapters and it would be good to recap what we have learned and highlight the points of importance before we continue onward.

Foundational checks

There are three foundational topology checks to pay attention to first and foremost. The other topology rules will be hard to check if these first three conditions are not met:

- All of the faces on the model are quads: they each have four vertices and four edges
- There is no duplicate geometry
- All of the face normals are oriented in the same direction so that you cannot see the back of the faces

Geometric topology rules

When all of the preceding checks are satisfied, we can look at our first set of topology rules. These are the rules that generally direct you on how to lay out the topology on your mesh, as described in *Chapter 2*:

- All of the loops on a grid need to terminate into themselves, or into the void
- None of the loops can overlap themselves
- A loop cannot spiral around the mesh

These are the **geometric topology rules** because they pertain to just the shape of a model.

Deformation topology rules

Next, let us revise the topology rules we learned about in *Chapter 3*, which are responsible for ensuring good deformations:

- The edges of our grid need to be parallel to the axis of deformation
- The axis of twisting should be in line with the edge flow
- You should always put at least one row of faces between a pole and the specific areas of deformation

Because these rules apply to deformation, they are called the **deformation topology rules**.

Now that we have those rules refreshed, we can start on the retopology. When approaching retopology, you always want to start in the areas of detail first, the areas where you really need it to look in a certain way. For instance, the head has more detail than the chest. On the head, the face has more detail than the back of the head.

On the face, there are eyes, a nose, and a mouth. So, on a head, you would start with those individual features, then connect them to make the face, and then fill out the rest of the head. This way, we can make sure we have all of the details we need where we need them.

To start on our retopology of the head, we need to lay meshes over each other, and for this, we need to use snapping, which we covered in *Chapter 2*.

How to use snapping

Snapping allows us to snap any mesh selection to a face, edge, or vertex. To enable snapping, press the magnet at the top center of the viewport, as shown in *Figure 5.3*:

Figure 5.3 – The magnet icon, which enables snapping

By default, it is set to snapping to **Increment**. We want to change this to snap to faces. To change this setting, we need to press the arrow next to the magnet, and a menu like the one shown in *Figure 5.4* should pop up.

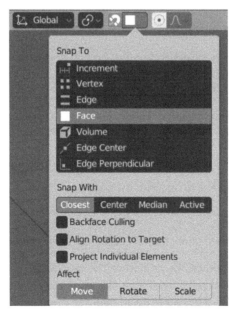

Figure 5.4 – Menu to change snapping default settings

Now, if we select a vertex in a separate mesh, we can snap it onto one of the faces of another mesh. With that, we should be able to start on our retopology.

Retopology of the face

The model we will be using to learn how to apply good topology rules can be seen in *Figure 5.5*:

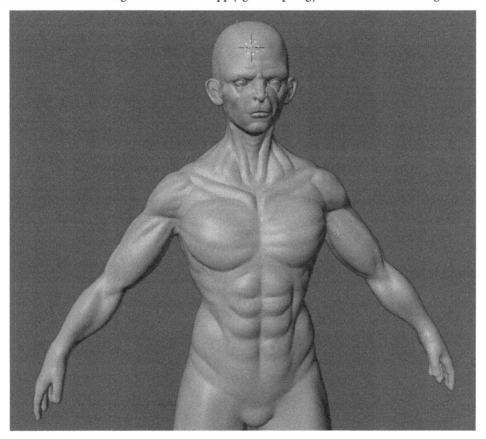

Figure 5.5 – Full body of the model to be used for retopology

The model has a unique face, which should give us an opportunity to stray from the topology guide and allow us to think about our topology beyond just copying from the reference. As said before, we want to start with the head because this is the area on the body with the most detail. On the head, we will start with the face for the same reason. In *Figure 5.6*, we can see the face a bit more clearly:

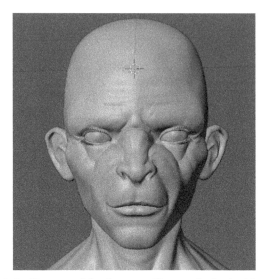

Figure 5.6 – Close-up of the face we will retopologize in this chapter

As you can see in the preceding figure, the face has many features to take into account including the mouth, nose, eyes, and ears. It is generally good practice to start by laying loops of vertices around any holes in the model.

Defining loops

Let us put a row of vertices around all the holes on the face including the eyes, nostrils, and mouth. In *Figure 5.7*, the loops of edges are all laid out:

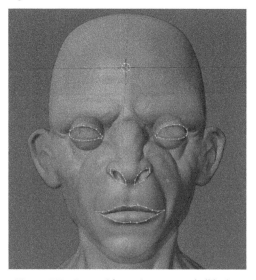

Figure 5.7 – Face with vertices around all holes

To add these loops, I first added a plane, then deleted all but one of the vertices, and I extruded the vertices using the *E* hotkey until I had all of the cavities surrounded by a loop of edges. Do not worry too much about the number of vertices at this point, as we will change them as we work. Just try to add the lowest number that feels natural. Next, we will mark all of the major creases by doing the same thing. When you are done, it should look something like *Figure 5.8*:

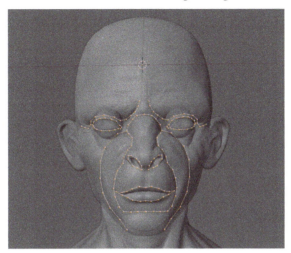

Figure 5.8 – Face with vertices around all creases, valleys, and holes

Now that we have the creases and valleys marked, we can look at the peaks/edges that are most prominent. *Figure 5.9* shows you what this model looks like after marking those edges:

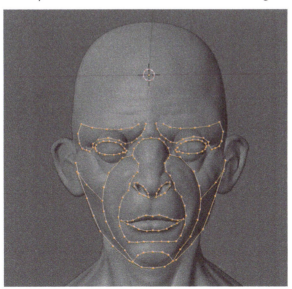

Figure 5.9 – Face with vertices around edges, creases, valleys, and holes

Now we can start to join these edges together. Remember, we are still trying to work from the areas of detail first, so do not be afraid to add more vertices to the areas that may not need them to accommodate the areas that do need those details in order to ensure a clean joining between the guiding edges.

Joining the edges

The place with the finest detail on this face is probably the eyes, so we can start there. In *Figure 5.10*, there is a closer view of the eye area:

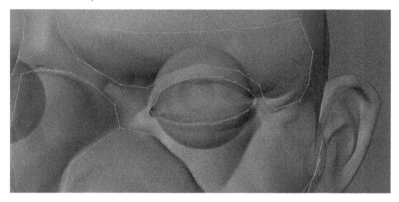

Figure 5.10 – Close-up of the eye area with vertices

It is plain to see that the surrounding edges on the brow and the start of the cheek have far too few vertices to actually connect to the ones that we laid out on the eyelid. That is no problem because it is easy to add more. The loop cut tool is a great way to add more vertices to an edge. Just select the edge you want to modify, press *Ctrl + R*, and scroll up on the mouse wheel to add more vertices. After adding the vertices and snapping them to the face, it should look like *Figure 5.11*.

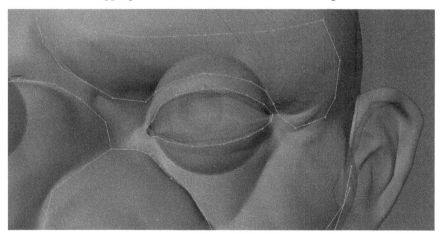

Figure 5.11 – Close-up of the eye with more vertices snapped to the face

Notice how the vertices are positioned to line up with each other as well. This will make it easier to join the faces together. To start joining the edges together, select two that line up nicely with each other, then press *F*. *Figure 5.12* shows the first face that was made to connect the two strings of vertices:

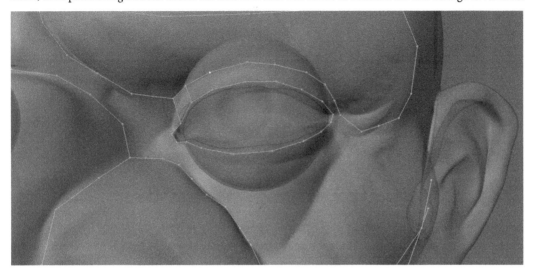

Figure 5.12 – First face connecting two strings of vertices

Next, we just need to fill in all of the joining edges around the eye. Now you should have a full loop of faces going around the eye, as in *Figure 5.13*:

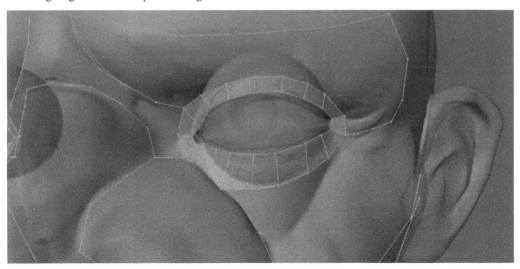

Figure 5.13 – Full loop of faces around the eye

We can use *Ctrl + R* to check our topology, and that will show us that this loop follows all of our topology rules. From here, we can move on to the nose. We can get a closer look at the nose in *Figure 5.14*.

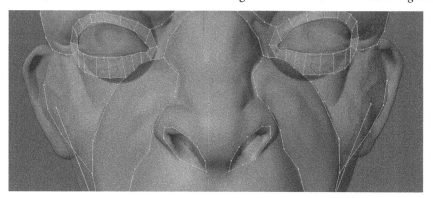

Figure 5.14 – A close-up of the nose with the first set of vertices

This shape is far more complex than the eye, so we should mark all of the edges and creases in this more focused section of the face. There are a few edges running down the bridge of the nose, and some creases around the nostrils that we have not marked as well. These are marked in *Figure 5.15*.

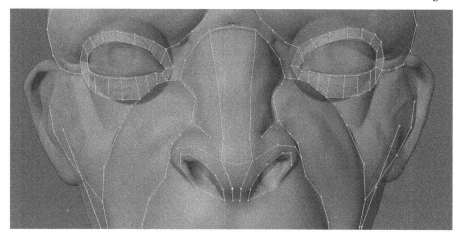

Figure 5.15 – The nose and nostrils with more detailed vertices

Notice how the number of vertices mostly line up with the edges next to them that were there previously. It is helpful to think ahead when laying out these edges because we are likely going to join them to the neighboring edges. With our new edges laid out, we can start placing our faces. First, we will start with the faces around the nose like in *Figure 5.16*.

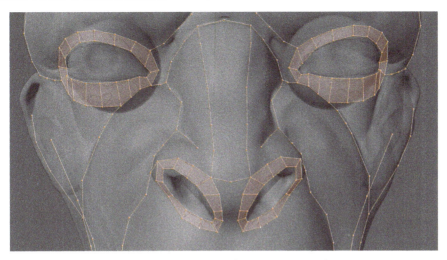

Figure 5.16 – Nose with faces around nostrils

With those loops done, we can work our way back to the bridge of the nose. All we have to do is join the edges that we have together, just like we did with the eyes. When you get back to the bridge of the nose, it will look like *Figure 5.17*:

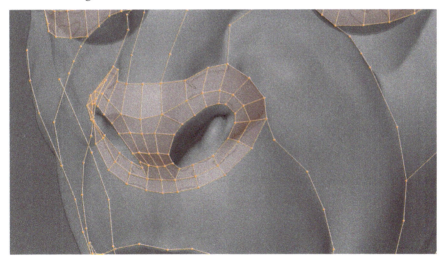

Figure 5.17 – Nose with additional faces over the bridge

One big thing to notice here is the loop going around the nostrils. Because the nose is an extrusion from the relatively flat face, it should have a loop going around it just like our examples in the *How should grids intersect?* section in *Chapter 2*, when we extruded a face from a plane. You can see the two loops next to each other in *Figure 5.18*:

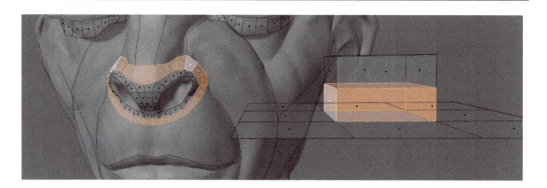

Figure 5.18 – The two adjacent nose loops

Now, we can do the same thing with the bridge of the nose, connecting all of the edges to form a clean grid:

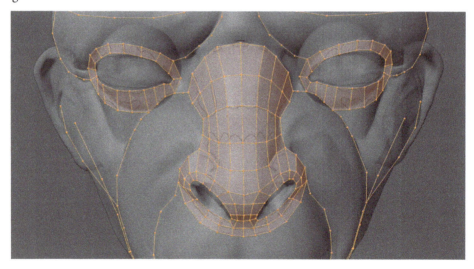

Figure 5.19 – Forming a clean grid on the nose

With that done, you can run your loop cuts through all of the faces and find that they all follow our topology rules. With the nose done, there is one last major feature on the face, and that is the mouth. This is actually handled in a manner very similar to the eyes. Let us take a closer look at where we left the mouth in *Figure 5.20*.

Figure 5.20 – Close-up of the mouth

With the mouth, we are just going to start by extruding the mesh on either side of the mouth loop by pressing *E* and snapping it to the faces. We can see the faces in *Figure 5.21*:

Figure 5.21 – Mouth with faces snapped over it

Now, we can take a step back and look at our vaguer major features. These are areas that are not as obvious as a hole or extrusion but still have a defined shape. A good area to start would be the selected edges in *Figure 5.22*:

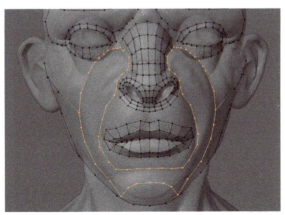

Figure 5.22 – Close-up of the next area to work on around the face and mouth

This part should be relatively simple, as all of the shapes seem to be forming a continuous loop around the nose and mouth. *Figure 5.23* shows this area all filled in:

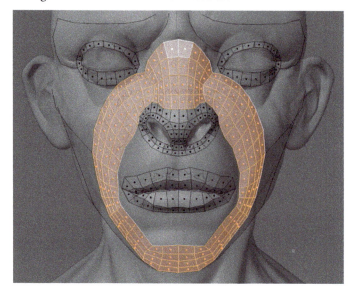

Figure 5.23 – The area around the face and mouth filled in with a looping grid

You might notice that the loop we made also flows up into the faces as well and has a very simple grid topology. This section also passes all of our topology rules. Next, we can do some work on the cheek in *Figure 5.24*.

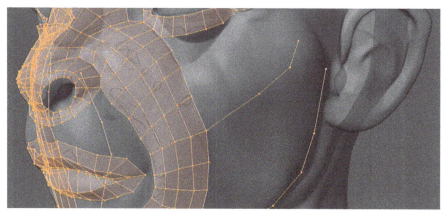

Figure 5.24 – Close-up of the cheek with guiding vertices

This topology should be even easier for us. We just need to extrude this face on either side of the guiding edge as well. Make sure to fill in the faces so that they roughly match the size of the rest of the quads on the face. You can see this in *Figure 5.25*.

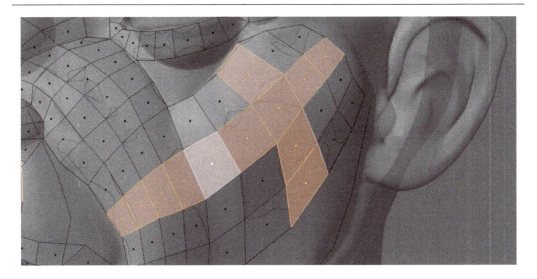

Figure 5.25 – Cheek filled with faces

This also follows our topology rules as it is just a simple grid deformed on the face of our character. With the cheeks done, there is not much left to fill out. *Figure 5.26* shows us what is left:

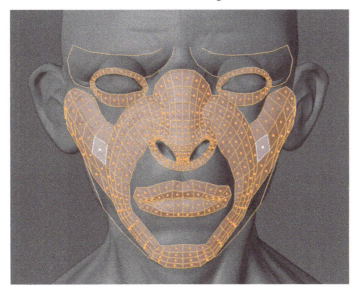

Figure 5.26 – Areas left on the face to retopologize

The brows are the last major shape we need to focus on before filling in all of the spaces in between. *Figure 5.27* gives us a closer look at what we are dealing with.

Figure 5.27 – Close-up of eyebrow

The eyebrows look a lot trickier, but really, they are not much different from the other areas that we have looked at. All we are going to try to do is fit a simple grid into this space without violating our topology rules. It can be helpful to run a string of edges through the area that you are trying to fill to get a better idea of what path the faces want to naturally follow. *Figure 5.28* shows this intermediate step.

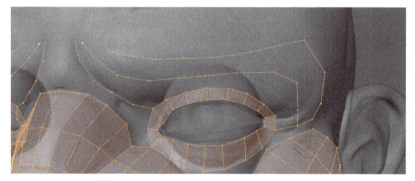

Figure 5.28 – Eyebrow with guiding edges

With the guiding edges in place, we can add our faces. It is the same technique as before: we just join the gaps with faces and adjust the vertices on the edges to line up properly. You can see the result in *Figure 5.29*.

Figure 5.29 – Eyebrow guiding edges filled with faces

With that, all of the major shapes on the face have been defined. In *Figure 5.30*, we can take a look at all of the sections together.

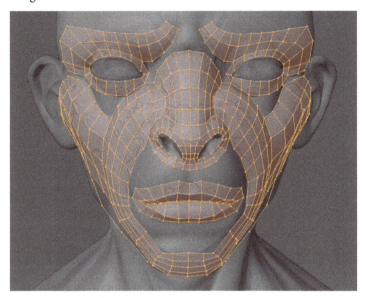

Figure 5.30 – Face with major shapes defined

Because all of our sections up until this point have followed all the topology rules, connecting the sections should be easy so long as we pay attention to our loops.

Joining the sections

The mouth has the most work, so we should start there. First, we will connect the two edges from *Figure 5.31*.

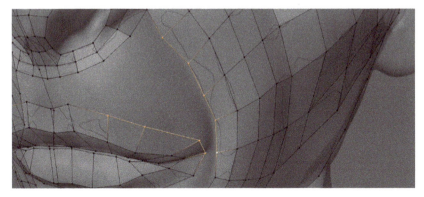

Figure 5.31 – The two edges on the face that will be connected first

Another guiding loop would definitely be helpful here. You can see this in *Figure 5.32*.

Figure 5.32 – Face with an additional guiding loop on the lip

Now we can start to join those faces. *Figure 5.33* shows these edges connected:

Figure 5.33 – Face with guiding edges connected

This connection does not break any of our rules, so now we just have to join the area between the nose and the mouth. First, we join the nose edges into the guide edges, as shown in *Figure 5.34*.

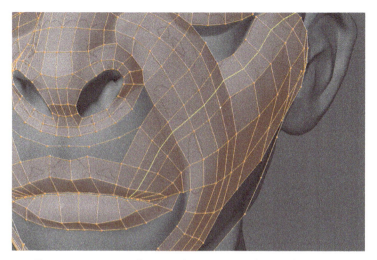

Figure 5.34 – Face with nose edges connected to guiding edges

And then finally, join the lip to the guide edges as well. Because the curvatures of the chin and the bottom of the mouth were so different, there are two guiding lines defining their curves. These are shown in *Figure 5.35* along with the finished upper lip.

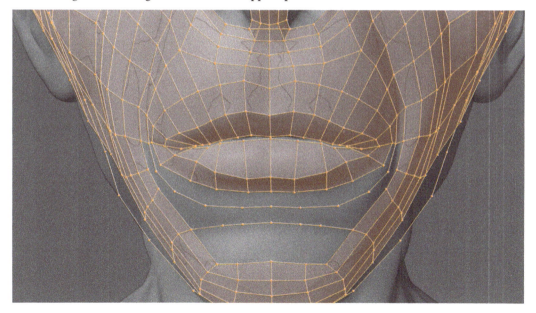

Figure 5.35 – Face showing areas remaining to be joined under the mouth

We are going to join these areas to the guiding edges just like we did on the top of the mouth. This can be seen in *Figure 5.36*.

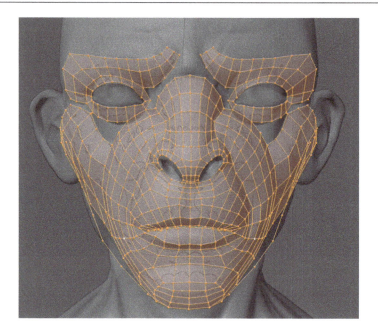

Figure 5.36 – Face with most edges connected

Now, we have the eye area in *Figure 5.37* to sort out.

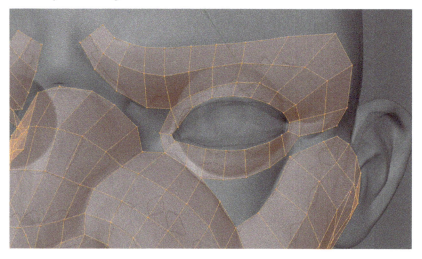

Figure 5.37 – The remaining eye area

This area is pretty messy and is going to be a bit more difficult to sort out. We will start with the easy part on the left. All we really have to do is join the faces directly to each other, and make sure they follow our topology rules. This is shown in *Figure 5.38*.

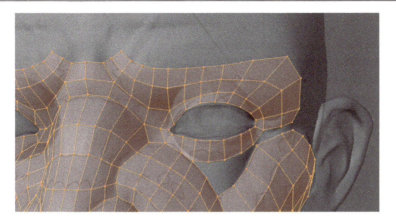

Figure 5.38 – Eye with edges joined on the left corner

Now we have to address the right side. The thing that makes this difficult is that this area is very triangular, and that is very much against our rules. To start, we are going to merge the vertices on the right corner of the eye and create a guide edge for the lower part, as shown in *Figure 5.39*.

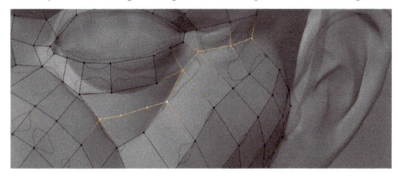

Figure 5.39 – Close-up of eye with guide edge

Next, we just have to join our edges together, and it should look like *Figure 5.40*.

Figure 5.40 – Eye with all of the faces filled in

With the major sections done, there are a few housekeeping things we can do. Generally, it is a good idea to try and terminate our face loops into themselves, or into the void as often as possible; this reduces the opportunities for the mesh to overlap itself, or to not line up properly. In *Figure 5.41*, you might see what the issue could be highlighted by a loop cut.

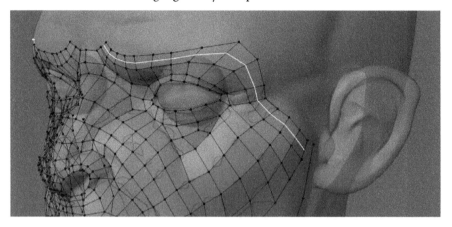

Figure 5.41 – Loop cut showing the potential issue

Because these edges are not going to terminate into any of the holes on the face, the highlighted edge is going to have to terminate into itself. As it is right now, we would have to keep track of the two ends of this loop of faces until we finally get them to loop back into themselves. If we join them like in *Figure 5.42*, we only have to manage one end.

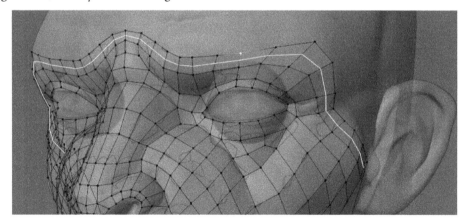

Figure 5.42 – Loop of faces joined at the top

We are going to repeat this on the chin, terminating the selected part in *Figure 5.43* back into itself:

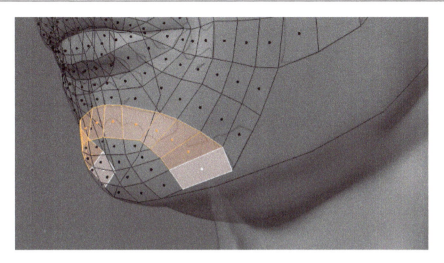

Figure 5.43 – Selected chin loop that needs to be terminated into itself

And in *Figure 5.44*, you can see the final loop:

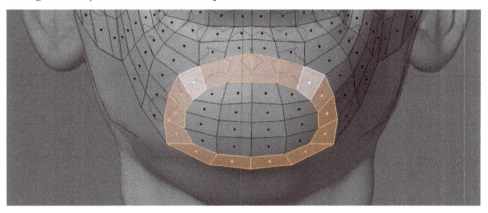

Figure 5.44 – Final loop on the chin terminating into itself

We only have the jaw left on the face, so we are going to fill that area next. In *Figure 5.45*, there is a closer view of his jaw with a guiding edge going through it:

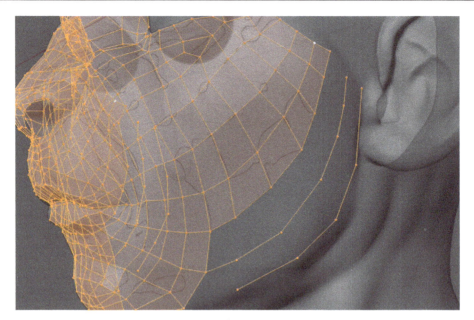

Figure 5.45 – Close-up of the jaw with guiding edges

And with the faces filled, it should look like *Figure 5.46*:

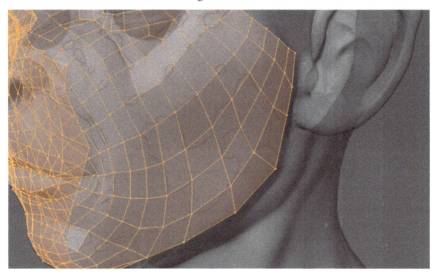

Figure 5.46 – Jaw with faces filled

And with that, we are done with the face. We can run loop cuts all around the face to find that all of the loops follow our rules. Now, we can take a look at some of the work we have done and the loops that make up the face. *Figure 5.47* shows us all of the sections that we had originally laid out:

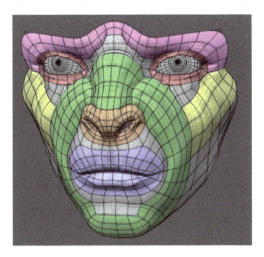

Figure 5.47 – A view of all the sections on the face

If you compare it to the original reference at the beginning of this chapter, you will notice that all of these colored sections look more or less the same. That is because a humanoid face will almost always have a nose, mouth, and eyes. When you model those, all that is left is joining them together. The white parts are the areas that are most likely to be different, as those are the areas that connect the features. That is why we always focus on the areas of detail first, making sure those areas follow our topology rules so that it is easy to join them together. The last difficult part of our head is the ear, and this is usually the most painful part of the head.

Retopology of the ear

Figure 5.48 shows us the ear we will be working on:

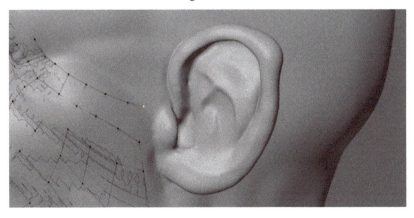

Figure 5.48 – Close-up of the ear

As we did for the face, we are going to map out the peaks and valleys, as shown in *Figure 5.49*:

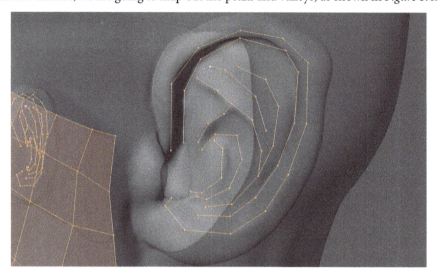

Figure 5.49 – Peaks and valleys defined with vertices on the ear

Then comes the agonizing task of joining the faces in a way that does not break topology rules. Unfortunately, there are not too many tricks here other than trial and error. Many 3D artists tend to use the ear to hide their messy topology. This is acceptable because the ear usually does not deform much, and is very easy to UV unwrap, even with bad topology. The level of detail in an ear, and its shape, can also radically change the topology required. For this reason, it is usually not worth potentially sinking hours into perfect ear topology. This ear ended up looking like *Figure 5.50*.

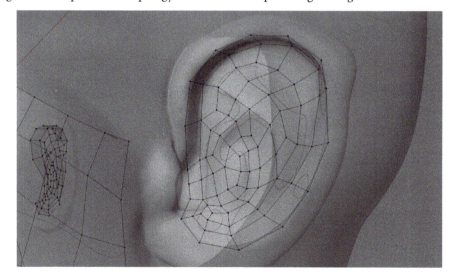

Figure 5.50 – Ear covered with faces

While it may not look like it, this ear actually follows all of our topology rules. With the inner part laid out, we can start to move to the outer part of the ear. Once you have worked your way all the way around the ear so that you have a ring of vertices surrounding it, as in *Figure 5.51*, we can start connecting it to the face.

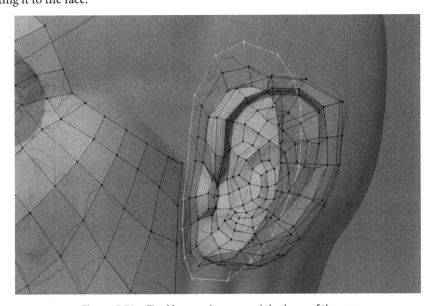

Figure 5.51 – Final loop going around the base of the ear

By adding the pole in *Figure 5.52*, we are able to split the faces to give us more geometry to join to the ear:

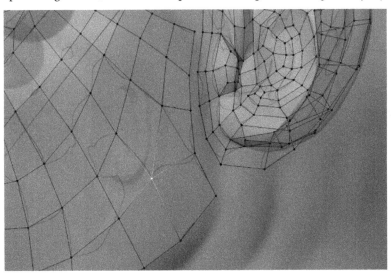

Figure 5.52 – Pole added to improve ear geometry and ease of joining

You could also loop-cut the face to add more geometry, but in this case, it would have made the face geometry too dense. Now that the vertices match up, we can merge the face to the ear, as in *Figure 5.53*.

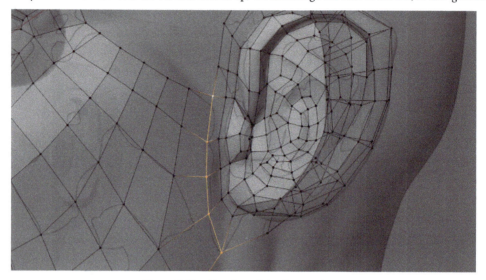

Figure 5.53 – Merging of the face and ear

With the ear done and connected to the face, we can start on the rest of the head.

Retopology of the back of the head

To begin with, we are going to join more of the ear to the head to get a flat line going across the two, as shown in *Figure 5.54*:

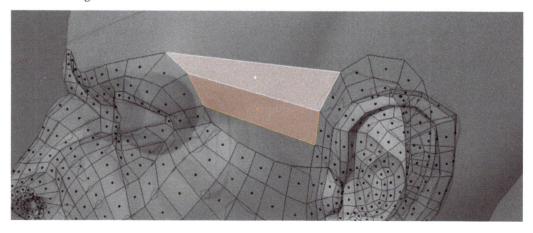

Figure 5.54 – Joining more of the ear to the head

Then we are going to loop-cut those faces to join into the rest of the face to match *Figure 5.55*:

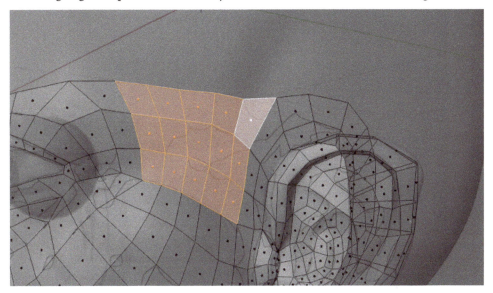

Figure 5.55 – Looping faces between the ear and face

Next, we are going to work toward the back of the head. Like the ear, the back of the head is not very important, so there is not too much to mess up here. To start, we are going to extrude the edges at the top of the area we just merged, as shown in *Figure 5.56*.

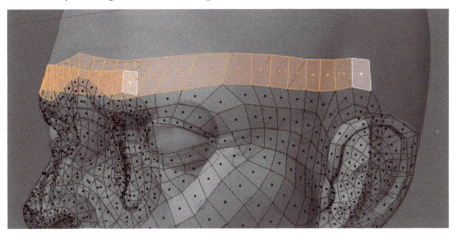

Figure 5.56 – Face with top edges extruded

Then, continue that loop to the back of the head, and add another loop across the top of the head, as in *Figure 5.57*.

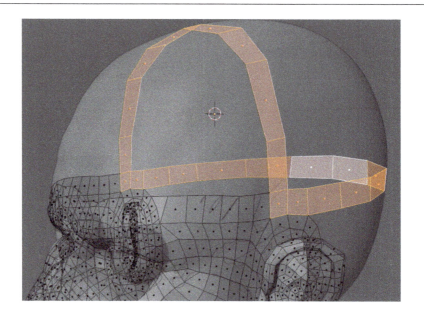

Figure 5.57 – Adding guiding loops to the top of the head

These will act as our guides as we fill the areas in between, as you can see in *Figure 5.58*:

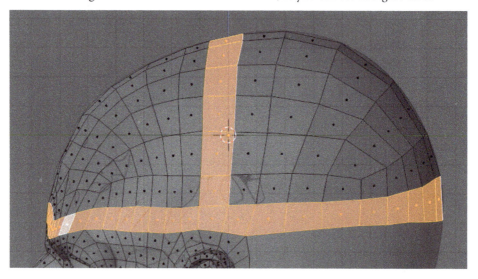

Figure 5.58 – A view of the guiding loops on and around the top of the head

And with that, the head portion of the model is done. We do not want to go too far down the back of the head because we may need to change things depending on how the head connects to the body.

Summary

In this chapter, you learned what retopology is. By now, you should know how to use snapping and how to retopologize a face, ear, and head.

This was a difficult chapter, far less direct than the previous ones. The most important thing to take away from this chapter is the process we used to break down the detailed sections and then bridge them together using our guiding edges. With the head out of the way, the most difficult part of the body is done. Now, we can take a step back and look at the rest of the body.

The next chapter will handle the rest of the body, starting with the hand. This is certainly a close second in difficulty to the head, but with our new knowledge from this chapter, we will confidently complete the task.

6
Topology on a Humanoid Body

With the head done in the previous chapter, we can move on to the rest of the body. On the head, we began working on the most detailed part, which is the face. We broke down the face into even more detailed areas, such as the nose, mouth, and eyes. Once we finished these areas, we only had to focus on connecting them and then filling out the less important parts of the head.

Next, we are going to use this same idea to approach the body. We will start with the detailed areas, and then work out from there. The hand is usually the most detailed part of the body, so that will be the section of the body we will start our work from. Next, we will look at the shoulder, and connect it to the neck. Then, we will move on to the hip and crotch area. Finally, we will join them all together.

In this chapter, we will be learning about the following subjects:

- How to retopologize a hand
- How to retopologize shoulders
- How to retopologize hips
- Connecting the body parts together

How to retopologize hands

When approaching hands, we need to break them down into parts. Thankfully, with the hand, it is not too difficult, as they are very naturally broken up into segments. The fingers and the palm make up the detailed areas of the hand. This makes it easy to start, but it also complicates joining the sections together. Ideally, you would have the detailed areas separate from each other. This makes it much easier to connect them because the area in between is less important. Because these detailed areas are right next to each other, we need to be more creative with how we join them together to ensure good and even topology. You can see our hand in *Figure 6.1*.

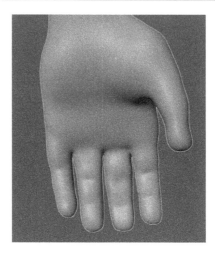

Figure 6.1 – Close up of a hand

On the hand, the fingers are a good place to start. The tip of the finger is usually the most detailed part of the finger, especially if the character has fingernails. In this case, our character does not, but the tips of the fingers are still the most detailed part of the fingers, so we will start there. *Figure 6.2* shows us the tips of the fingers, and how we started them.

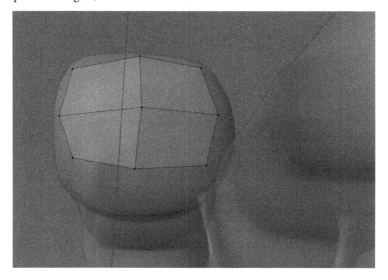

Figure 6.2 – Close up of a fingertip with preliminary vertices

A simple grid was snapped to the tip of this index finger. We want to keep the mesh as minimal as possible to keep the rest of the hand at a reasonable quad count. In *Figure 6.3*, we extruded the mesh inward by selecting all of the faces and pressing *E*.

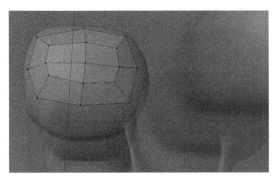

Figure 6.3 – Faces extruded and snapped to the fingertip

Extruding the mesh inward like this and snapping it to the model allows us to add a little more detail without affecting other parts of the model. The next step is to extrude the border of this plane down to the first knuckle and snap it in place, as in *Figure 6.4*.

Figure 6.4 – The border of the plane snapped in place at the first knuckle

This makes it easier to add loop cuts afterward because Blender will interpolate the loop cut between one end and the other. *Figure 6.5* shows this in a little more detail:

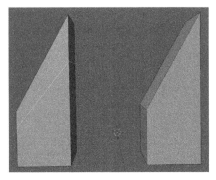

Figure 6.5 – Interpolation of the loop cuts

These are identical shapes with different loop cuts applied to them. The one on the left has a loop cut closer to the top, and the shape on the right has a loop cut closer to the base. As the loops get closer to the top, they will look more like the top loop, and as they get closer to the bottom, they will look more like the bottom. You can see a shape with multiple loop cuts in *Figure 6.6* to help illustrate this further.

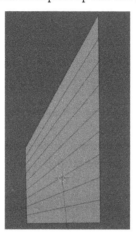

Figure 6.6 – Shape with multiple loop cuts showing interpolation of the loops

We will do the same thing with the finger after we extrude it all the way to the base of the finger. First, we need to extrude it to the next knuckle, then to the base, and snap both of those extrusions into place, as in *Figure 6.7*.

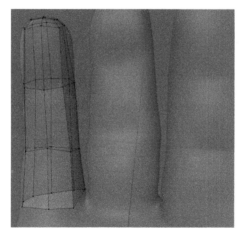

Figure 6.7 – Finger with extrusions snapped into place

This is a good place to duplicate this mesh to the other fingers. To duplicate the mesh, select the whole mesh, and press *Ctrl + D*. After the mesh is duplicated, move it to the next finger and snap it in place. When you have done the four fingers other than the thumb, it should look like *Figure 6.8*.

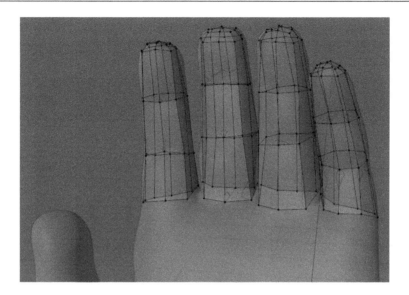

Figure 6.8 – Mesh duplicated and applied over all fingers except the thumb

With the fingers duplicated, we can add details such as the knuckles. For the knuckles, we will be using the same technique we used on the elbow joint in *Chapter 3* under the *Fixing the topology on an elbow joint* section. First, we will add loop cuts in between the joints of the finger, as shown in *Figure 6.9*.

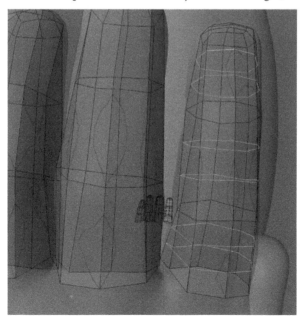

Figure 6.9 – Loop cuts added between finger joints

Do not worry about snapping these to the hand just yet. Instead, we are going to continue with the joint. To make the knuckles, just select the faces on the top and bottom of the edge loops selected, as shown in *Figure 6.10*.

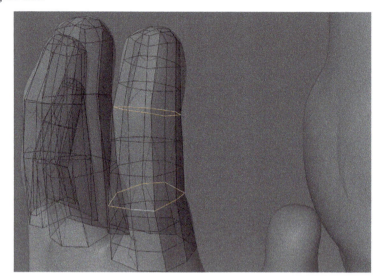

Figure 6.10 – Faces selected to mark the knuckles

When selecting the faces, stop your selection halfway around the back of the joint. It should look something like *Figure 6.11* when you are done.

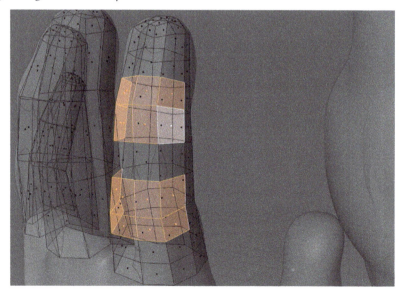

Figure 6.11 – Faces selected to inset into the knuckles

With these faces selected, we can inset the faces by pressing *I*. *Figure 6.12* shows the finger after the inset.

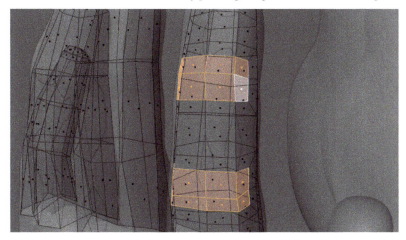

Figure 6.12 – Finger after being inset

Repeat this step for the other fingers as well. You can refrain from manually snapping these vertices to the hand as well. Because we copied the fingers and snapped them to the joints before, we can apply a shrink wrap modifier to snap these vertices for us. These are the steps to apply the modifier:

1. First, go to the **Modifier** tab and select the **Add Modifier** tab, as shown in *Figure 6.13*:

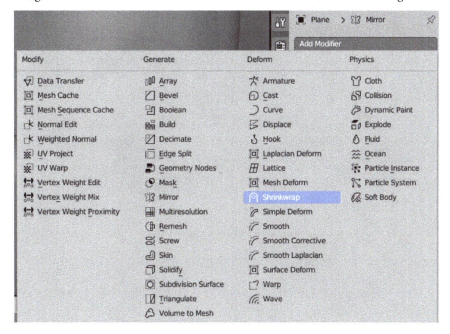

Figure 6.13 – Menu under the Add Modifier tab

2. After selecting the **Shrinkwrap** modifier from the **Add Modifier** tab, select the dense mesh that you have been snapping to manually.

3. To select the dense mesh, click on the **Target** option on the **Shrinkwrap** modifier. And in the dropdown, select the dense mesh. In this case, I have called the dense mesh Monkey Man. With the dense mesh selected as the target, the modifier should look like *Figure 6.14*.

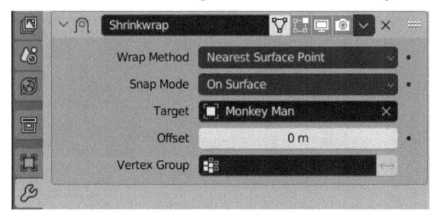

Figure 6.14 – Modifier with dense mesh selected

4. Finally, we can select the arrow pointing down on the modifier, and apply the modifier. You can see this in *Figure 6.15*:

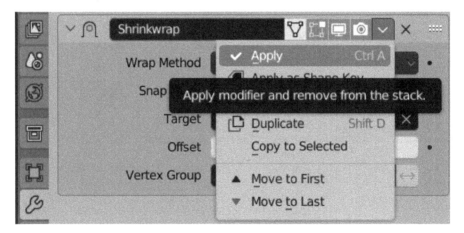

Figure 6.15 – Menu to apply the modifier

After smoothing the topology to match the shape of the knuckles a little better, *Figure 6.16* is what we are left with:

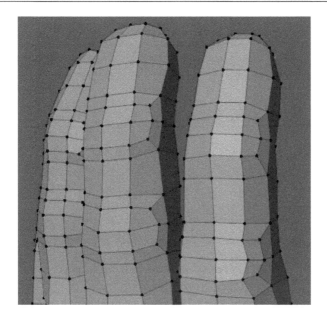

Figure 6.16 – Fingers with modifiers applied

Next, we can start on the palm. You could also start with the thumb, but it can be difficult to join things up. Using the thumb to accommodate the geometry usually makes things easier. So, as we did on the face, we need to outline the creases and peaks. *Figure 6.17* shows the simple outline.

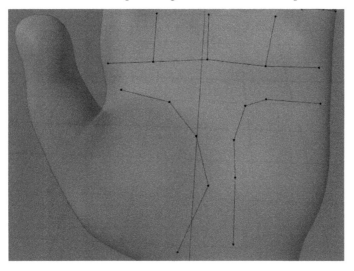

Figure 6.17 – Palm with creases and peaks outlined

With the lines laid out, we can fill them in with simple grids, as in *Figure 6.18*.

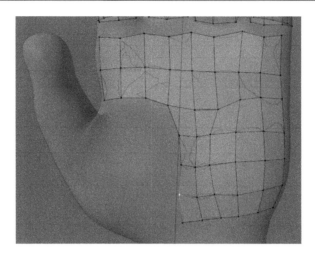

Figure 6.18 – Palm partly filled with simple grids

Now, we can flip our hand around and do the same thing with the back of the hand. In this case, it should be pretty easy. *Figure 6.19* shows the back of the hand with the sculpted mesh hidden.

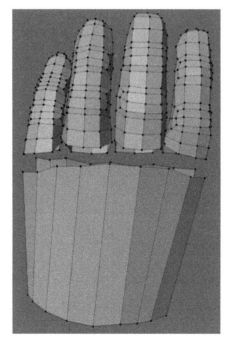

Figure 6.19 – The back of the hand with faces on it

Next, we join our fingers to the rest of our hand on the front and the back, as in *Figure 6.20*.

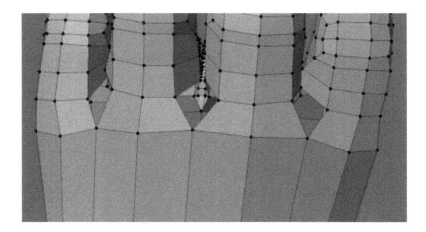

Figure 6.20 – Fingers joined to the rest of the hand

You may have noticed this left a few gaps between the fingers, but these are easily fixed. *Figure 6.21* shows the area filled in between the fingers.

Figure 6.21 – Area between the fingers filled

Repeat this joint for all of the fingers. That leaves the gap around the outside of the hand, and in particular, two tricky loops, one of which is highlighted in *Figure 6.22*.

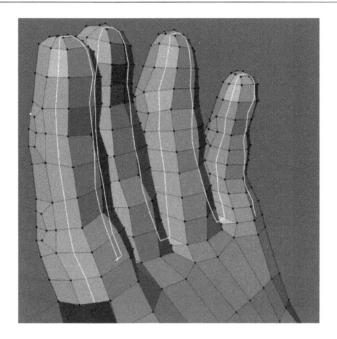

Figure 6.22 – Loop cut showing the flow of topology through the faces

All we have to do is join the sides together and prepare to extrude the thumb. *Figure 6.23* shows the mesh before extruding the thumb.

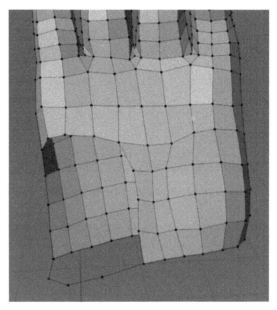

Figure 6.23 – Mesh before extruding the thumb

Next, we can select all of the vertices surrounding the thumb, and extrude them out by pressing *E*. Lastly, we just need to snap the vertices of the extrusion to the first joint of the thumb. But we have to be very careful when we join these faces together. *Figure 6.24* shows a highlighted loop going around the hand, and the completed thumb extrusion. We need to make sure that all of these loops connect back to themselves.

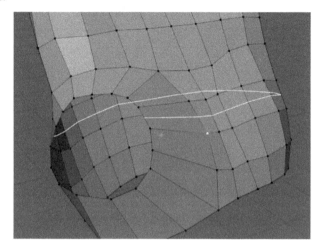

Figure 6.24 – Loop cut showing an example of a loop that needs to be connected to itself

It is helpful to join these together at this point, and then add loop cuts as needed to get it to match the shape of the thumb. You can see the loop completed in *Figure 6.25*.

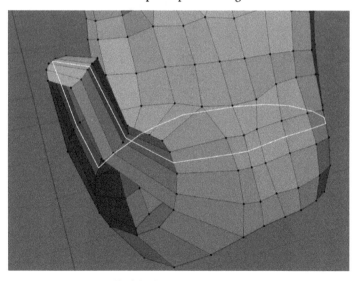

Figure 6.25 – All of the loops connected into themselves

After filling the remaining gaps from the other direction and snapping the vertices, it should look like *Figure 6.26*.

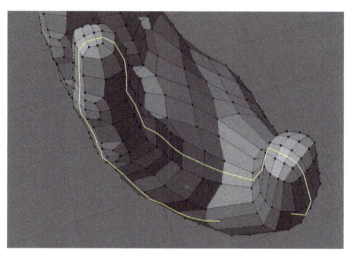

Figure 6.26 – The flow of topology on the snapped faces

The hand also follows our topology rules when we send loop cuts in this direction too. Now, all we have to do is add some loop cuts around it and inset our knuckle as we did on the other fingers. *Figure 6.27* illustrates the finished hand.

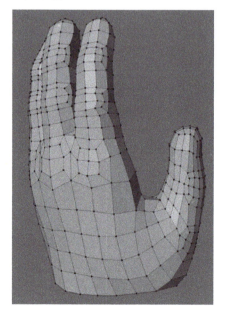

Figure 6.27 – The completely retopologized hand

All of the sections of the hand follow our topology rules, so we should not have to worry about the hand anymore. Thankfully, it is all downhill from here. With the hand done, we have to figure out how we are going to connect it to the rest of the body.

How to retopologize shoulders

Before getting up to the shoulder, we have to extrude our wrist up to the elbow, then to the shoulder, just like our fingers. You can see the arms extruded and with loop cuts in *Figure 6.28*.

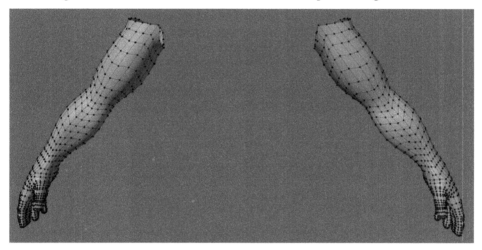

Figure 6.28 – Arms with loop cuts

Now we can start on our shoulder. Just like our face and hand, we are going to start by laying out some guiding lines. *Figure 6.29* shows the lines laid out for the shoulder.

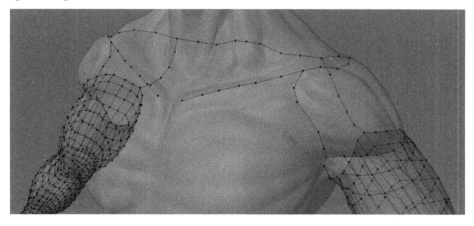

Figure 6.29 – Shoulder with guiding lines

And like the areas of detail before, we just need to fill these in, as in *Figure 6.30*.

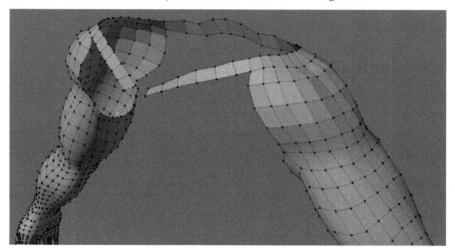

Figure 6.30 – Shoulder filled in with faces

With the shoulders roughed out, we can bring the neck down to the shoulders. For this model, I simply adjusted the loops on the collarbone to match the loops on the neck and did a straight connection, as shown in *Figure 6.31*.

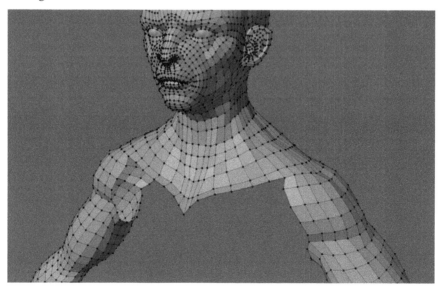

Figure 6.31 – Head, shoulders, and arms connected with edges

Now, we just have to work our way down the chest, and then do the lower body. But first, we should look down at the hips, and then focus on connecting the two parts of the body.

How to retopologize hips

The hips of the model will be done in a way that should seem familiar, as we have looked at a shape similar to a hip when we used the pant model in *Chapter 2* and *Chapter 4*. You can see the guiding lines in *Figure 6.32*.

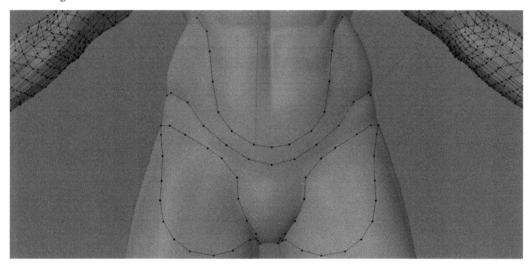

Figure 6.32 – Guiding lines across the hips

After connecting the guiding lines, the model should look something like *Figure 6.33*.

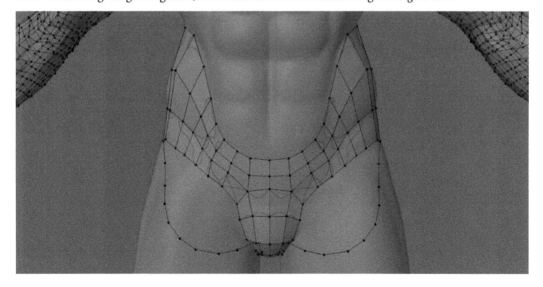

Figure 6.33 – Hips with guiding lines connected

With the front done, we can move on to the back. This will be a bit different from the front, but still not too difficult. You can see the guiding lines for the back in *Figure 6.34*.

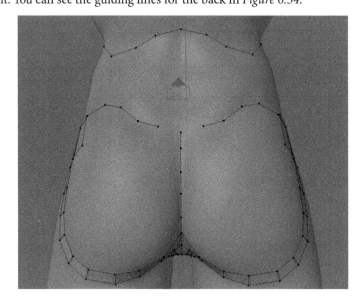

Figure 6.34 – Back with guiding lines

With everything connected, the back should look like *Figure 6.35*.

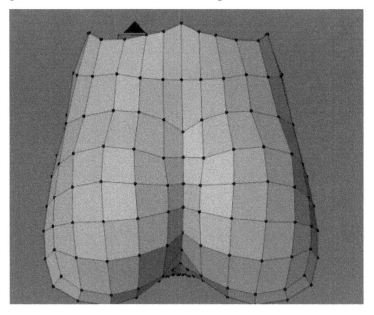

Figure 6.35 – Back with guiding lines connected

And with that, our hip is complete, and our model is just missing his torso now. You can see what we have left to do in *Figure 6.36*.

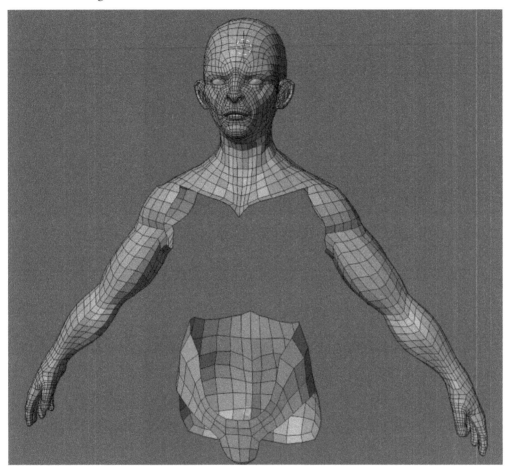

Figure 6.36 – Model with parts remaining to be retopologized

The gaps left over in the mesh will be filled in the next section to complete our torso.

Connecting the body parts together

For the chest, instead of just lines, we are going to use full faces. Because we have the sections already fully modeled, it can be helpful to visualize how the topology is going to react with full faces. *Figure 6.37* shows what this looks like.

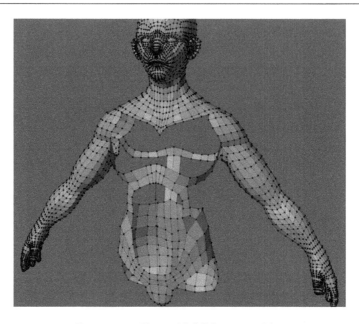

Figure 6.37 – Chest with full faces as guides

Again, these faces just outline the major shapes of the model, so that we can fill them in later. The filled-in model can be seen in *Figure 6.38*.

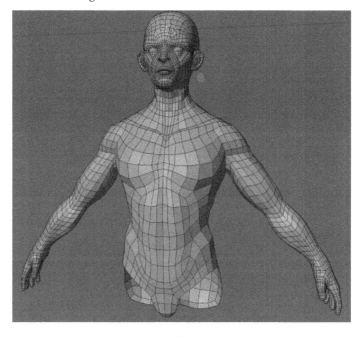

Figure 6.38 – The completely filled-in model without the legs

With that, we are almost done with the body. All that is left is to extrude the legs from the hip. All we have to do to achieve this is repeat what we did for the arms and finger:

1. Extrude the mesh to each joint.

2. Loop cut to get even quads.

3. Then, inset the faces on the outside of the joint.

After extruding out the legs and using a shrinkwrap modifier as we did on the arms and fingers, the retopology is done. We should be left with something like *Figure 6.39*.

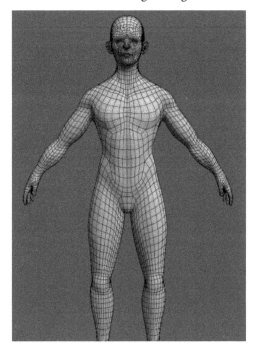

Figure 6.39 – The complete model of the body

Summary

In this chapter, we learned how to retopologize a hand, shoulder, and hip area. We also went through how interpolation works for loop cutting and learned how to connect detailed areas together.

By this point, you should have a comfortable understanding of the topology rules, and a good idea of what to do when approaching a new section of a mesh. While *Chapter 5* and this chapter focused on organic and flowing topology, *Chapter 7* will take a closer look at hard surface retopology using quads. In the next chapter, we will be confronted with topology situations that we would not ordinarily encounter when performing organic modeling.

Topology on a Hard Surface

Now that we have some experience with retopology on an organic shape like our monkey man, we can look at topology on a hard surface model. Thankfully, it is not too different from our normal modeling with one major exception: normals on hard surface models are far more important.

We are going to use the same techniques on a hard surface as we did on our organic model. We will start by breaking the model into areas of detail. If the model is already made of multiple parts, this can be a good place to start.

In this chapter, we will be exploring the following subjects:

- Normals on a hard surface
- Retopology of the grip of a blaster
- Retopology of the front shielding
- Retopology of the front grip
- Retopology of the barrel
- Retopology of the main body of the blaster

Normals on a hard surface

As we learned back in *Chapter 2*, under the *Understanding good topology using grids* section, the normal is the direction that the faces are facing. One feature we have not talked about yet is smooth shading. The shading shown in *Figure 7.1* is what we have experienced so far and this is called flat shading.

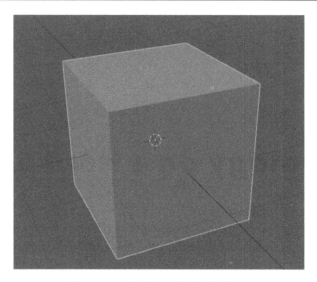

Figure 7.1 – A cube with flat shading

Flat shading shows the normals exactly as shown by the geometry. You can see every face individually and each face is separated by a sharp edge. **Smooth shading** averages the normals of all of the faces to smooth out the edges of intersecting faces. You can see that the shading is set to smooth in *Figure 7.2*.

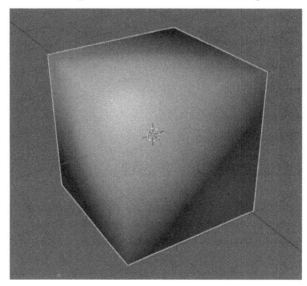

Figure 7.2 – Cube with smooth shading

You can change the shading mode of a model by going to the **Object** tab at the top of the viewport and selecting either **Shade Smooth** or **Shade Flat** as shown in *Figure 7.3*.

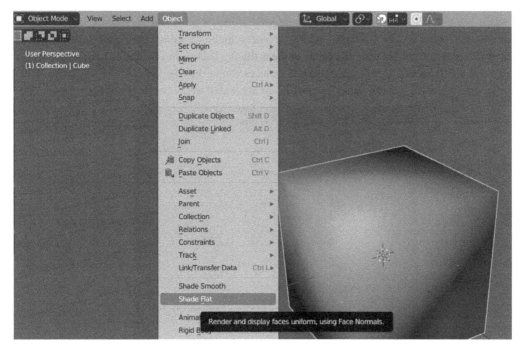

Figure 7.3 – The Object tab with smooth and flat shading options

Notice how it looks just fine on our organically shaped mesh, but as soon as we try and use smooth shading on something that is supposed to have sharp corners, it starts to look a little off. *Figure 7.4* shows a cube with smooth shading applied to it.

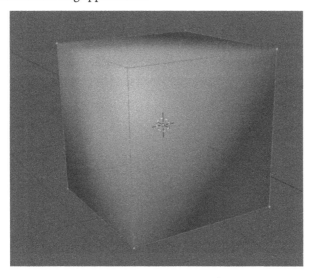

Figure 7.4 – Smooth shading in Edit Mode

As you can see, the shading is now trying to average the normals across a ninety-degree angle. This is causing some severe **shading artifacts**. Shading artifacts can be anything causing the final surface of the model to look off. In this case, the cause of the artificing is that the shading is having to smooth corners that are too sharp. If we were to add more faces where the edges are by beveling them, then the cube faces will start looking flatter as more levels are added to the bevel, as shown in *Figure 7.5*.

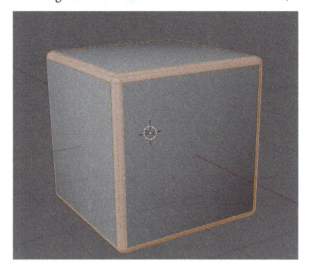

Figure 7.5 – A beveled cube

To bevel the edges, press *Ctrl + B*.

The normals are flatter on the main surface because now that there are more faces on the bevel the angle between each of the faces is more gradual, making the smoothing between each face far less extreme. This result can be good for some situations, but if you take a look at the corner close up like in *Figure 7.6*, the corner is still being smoothed a little too much.

Figure 7.6 – Smoothed beveled edges

A result that may be preferable would be something similar to *Figure 7.7*.

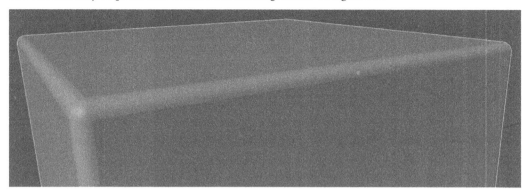

Figure 7.7 – Sharper beveled edges

This sharper bevel is caused by some sort of separation in the normals of the face to keep the edges sharp. There are a few different ways to do this, but we primarily are going to focus on two methods.

Auto Smooth

The first of these methods is by using **Auto Smooth**. **Auto Smooth** is a setting that allows you to decide what edges are smoothed based on the angle between their connected faces. The setting can be found in the **Normals** dropdown in the **Object Data Properties** tab shown in *Figure 7.8*.

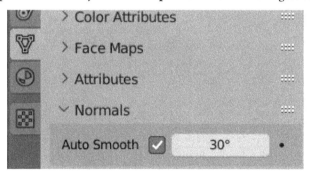

Figure 7.8 – The Normals tab in the Object Data Properties tab

If we enable this on our smoothed cube in the viewport, you will notice immediately our cube will get its sharp edges back. This is because the angle of all of the edges is 90 degrees and this is above the threshold in the **Auto Smooth** setting, which is set to 30 degrees by default. You can change this threshold in the **Auto Smooth** settings shown in *Figure 7.9*.

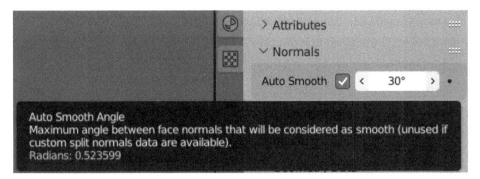

Figure 7.9 – Auto Smooth settings

This method is great if you are not planning on using beveled corners and if all of the corners you want to stay sharp are above the angle threshold. If this is not the case, the second method will allow us to manually mark the edges that we want to be sharp.

Mark Sharp

For this method, we need to go into Edit Mode again. This method also requires **Auto Smooth** to be enabled. You can set this to any angle you need for most of your edges, or set the threshold to 180 degrees to keep it from sharpening any of the edges. You can see a cube with **Auto Smooth** set to this in *Figure 7.10*.

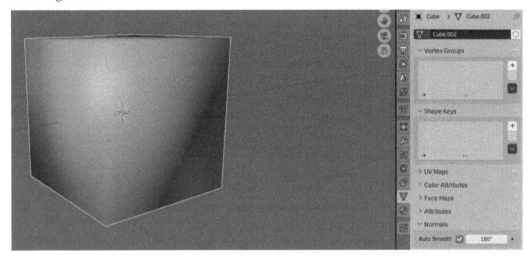

Figure 7.10 – Auto Smooth set to 180 degrees with smooth shading enabled

With **Auto Smooth** enabled and set to the desired angle, we can start manually marking the edges we want to sharpen in Edit Mode. To mark an edge as sharp, you need to do the following:

1. Select the edge you want to sharpen.

2. Press *Ctrl + E* to open the **Edge** menu.

3. Select the **Mark Sharp** option on the dropdown shown in *Figure 7.11*.

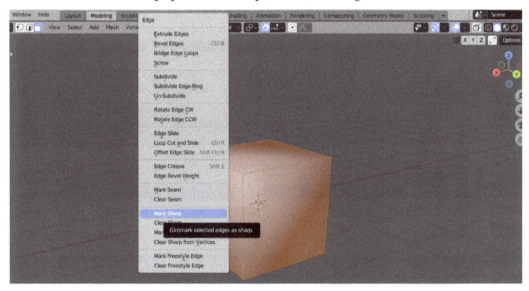

Figure 7.11 – The Edge menu

After marking the edge as sharp, there should be a blue highlight on the marked edge as in *Figure 7.12*.

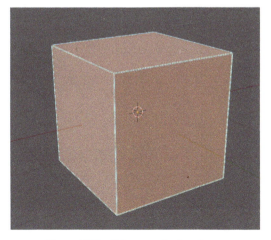

Figure 7.12 – Edges marked as sharp

Knowing how to mark our edges as sharp will make it much easier to visualize how our topology looks when retopologizing. With this new knowledge regarding normals, we are ready to start looking at the retopology of a hard surface model.

Retopology of the grip of a blaster

To practice topology on a hard surface model, we are going to be retopologizing the blaster shown in *Figure 7.13*.

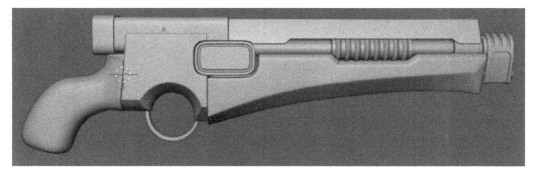

Figure 7.13 – The blaster

Like our previous methods, when we start, we want to break the model down into segments. In this case, we are going to start with the grip. *Figure 7.14* gives us a closer look at the grip.

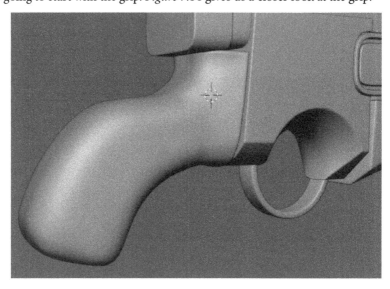

Figure 7.14 – The grip of the blaster

To start the retopology, we are going to snap a separate mesh onto the grip. Just like in our previous models, we are going to start by outlining the model with guiding vertices. First, we will put vertices on any creases or sharp edges. You can see an example of this in *Figure 7.15*.

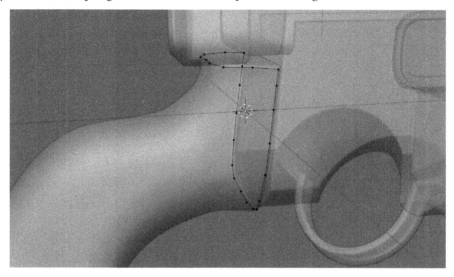

Figure 7.15 – Vertices placed on the edges of the grip

Next, because the grip is rather cylindrical, we are going to define the loops of the grip in a few increments as we did on the arms and fingers of our character. For this model, we ended up doing three loops as well as the loops of vertices that we already did around the edges at the top. *Figure 7.16* shows these loops.

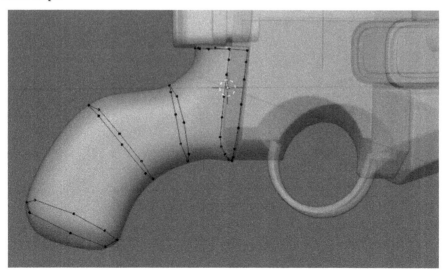

Figure 7.16 – Guiding edges on the grip

These loops do not need a lot of detail because we are going to use a shrinkwrap modifier to help us fill in any areas later. These loops are just there to get the flow of the topology right for the next step. Our next step is to join those simple loops together to get something like *Figure 7.17*.

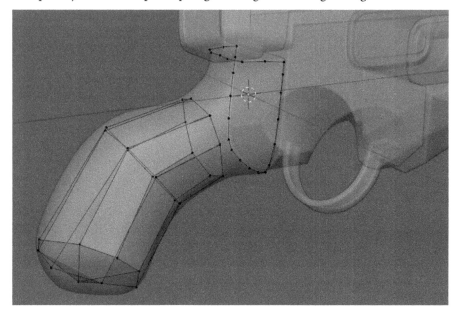

Figure 7.17 – Guiding edges filled in with faces

Now, we have a problem at the top of our grip. There are far more vertices on the right in *Figure 7.18* than there are on the left.

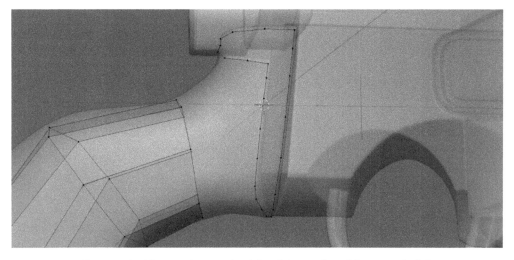

Figure 7.18 – More vertices on the right of the mesh and fewer on the left

To address this, we are going to start by extruding the vertices on the end into something similar to *Figure 7.19*.

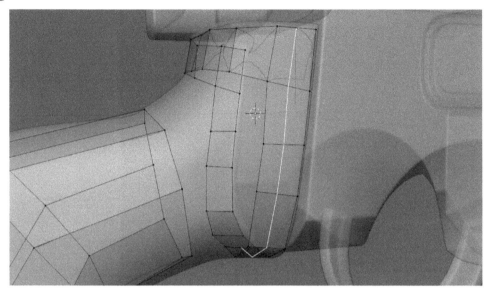

Figure 7.19 – Extruded vertices from the edge

Make sure that the edge loops going through these faces intersect like a normal grid. This is because we are trying to move the pole selected in *Figure 7.20* closer to the faces we need to connect it to in order to prevent stretching.

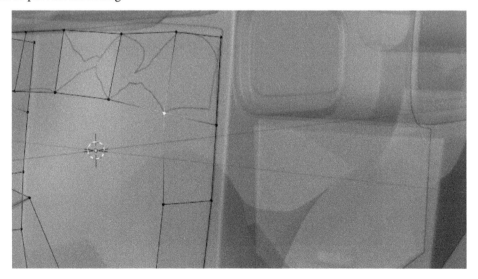

Figure 7.20 – Pole that will have five edges

With this pole now moved closer, we can loop cut the mesh on the left side to make it match with the right. Do not worry too much about snapping at the moment. When you have all of the vertices matching up on either side, you can finally connect it like in *Figure 7.21*.

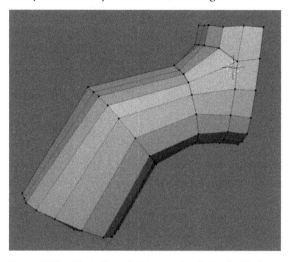

Figure 7.21 – The Left and right meshes joined with faces

Now, we can add loops to even out the topology. To add more than one loop, just do a normal loop cut with *Ctrl + R*, then scroll up on the mouse wheel. Then, we can apply a shrink wrap modifier in the modifier tab. On the shrink wrap modifier, make the target the object you are snapping to. Now, just apply the modifier. After the loop cuts and shrink wrap, it should look like *Figure 7.22*.

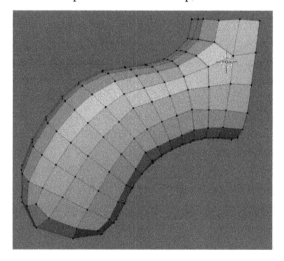

Figure 7.22 – Finished grip

With that, our grip is done. We can now start on the next part of our blaster, which is the front shielding.

Retopology of the front shielding

The next section we are working on is up at the front of the blaster. It is the angled plate that sits above the barrel illustrated in *Figure 7.23*.

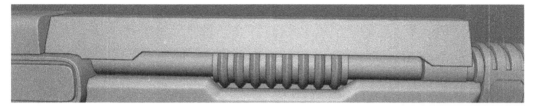

Figure 7.23 – The front shielding

Like we did on the grip, we are going to start by outlining the edges of the model. In this case, it is rather easy, as the shape is relatively simple. You can see the outline in *Figure 7.24*.

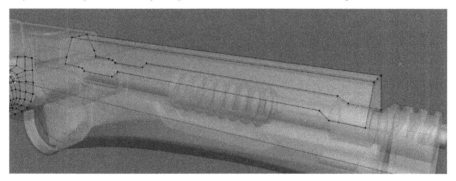

Figure 7.24 – Edges outlined with vertices

The important thing to think about in this case is how we are going to make the faces of the front shielding that are seen in *Figure 7.25*.

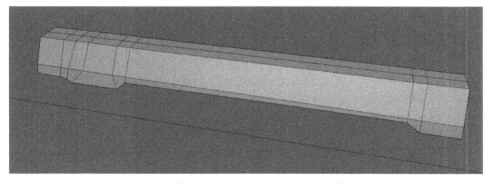

Figure 7.25 – Outlined vertices joined with faces

Here is the model after connecting all of the faces on it. This is the part of a deforming model where you would add more loop cuts in the middle to give it a consistent grid density. On a hard surface, however, there is no need to add more geometry because the model is not going to be deforming. All we have to worry about is the shading when we enable smooth shading. *Figure 7.26* shows us the model with smooth shading enabled.

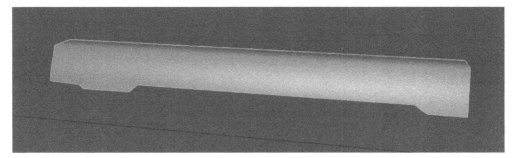

Figure 7.26 – The mesh with smooth shading applied

Now with smooth shading on, we cannot make out any of the edges of our model. The first thing when addressing this issue is to turn **Auto Smooth** on. For this model, we have a few curves that need to be smoothed, such as on the grip, and some subtle edges that need to be sharp. This means we cannot rely on **Auto Smooth** alone to get the desired shading. In this case, we are going to set the **Auto Smooth** angle to 60 degrees. This is a good threshold that usually sharpens only the edges that we want to sharpen. For the edges with an angle less than 60 degrees, we will have to mark them as sharp manually. *Figure 7.27* shows the model with **Auto Smooth** enabled and set to 60 degrees.

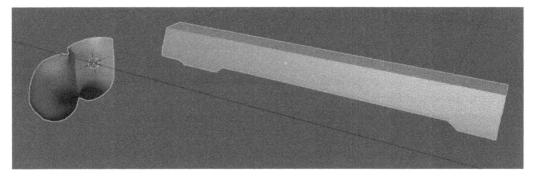

Figure 7.27 – The mesh with Auto Smooth set to 60 degrees.

This is a good improvement but there is still one small edge on the side of the shielding that we need to sharpen. To achieve this, go through the following steps:

1. Go into Edit Mode.
2. Select the edge you want to mark.

3. Press *Ctrl + E* to bring up the **Edge** menu.

4. Then, select **Mark Sharp.**

With the edges marked sharp, our model looks like *Figure 7.28*.

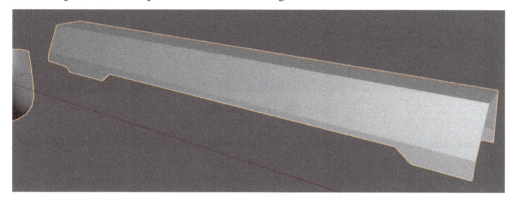

Figure 7.28 – Mesh with edges marked sharp

Now, all that we have to do is add some thickness to the shield. To do this, we just have to extrude the edges around the shield and snap them in place. You can see an example of this in *Figure 7.29*.

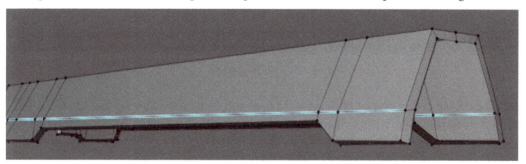

Figure 7.29 – Sharp edges in Edit Mode

With that, our shielding is done and we can move on to the next segment of our blaster, the front grip.

Retopology of the front grip

The next part of the blaster we are going to look at is the front grip. *Figure 7.30* shows a close-up of the front grip.

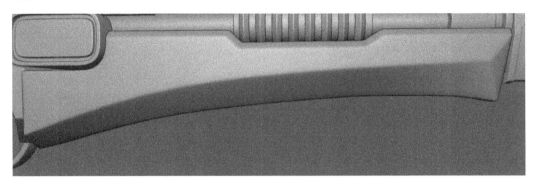

Figure 7.30 – The front grip

Again, we are going to lay out the lines on our edges and corners first. You can see the guiding edges in *Figure 7.31*.

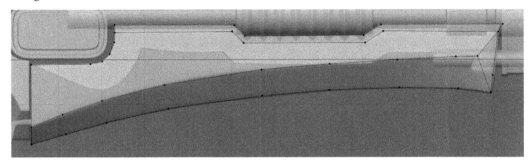

Figure 7.31 – Guiding edges

Next, we are going to start making the faces of the front grip. Most of the faces are going to be easy to place, but we should focus on the radius at the top left of the grip. You can see that there are far more vertices on the radius than the surrounding mesh. To address this, we just need to send the loops around the bottom of the grip. *Figure 7.32* shows the finished faces.

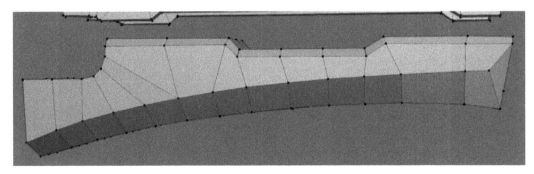

Figure 7.32 – Guiding edges joined with faces

After adding new geometry, you may need to smooth the model again. The smoothed faces can be seen in *Figure 7.33*.

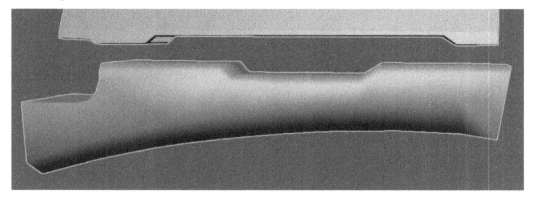

Figure 7.33 – Smooth shading applied

You can see that the auto-smoothing did not do as much as it did on the previous models, so this means we need to do a bit more manual edge marking. *Figure 7.34* illustrates the model after the edges have been marked.

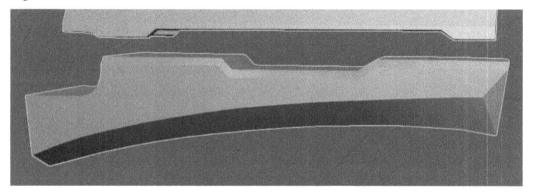

Figure 7.34 – Edges manually marked sharp

With that, the front grip is done and we can move on to the barrel of the blaster.

Retopology of the barrel

For the barrel, it will be helpful for us to hide the sections of the blaster that are obscuring it. To do this, simply select all of the parts of the model that you want to hide in Edit Mode, then press *H*. This will hide all of those parts of the mesh in Edit Mode only. To unhide them, just press *Alt + H*. In *Figure 7.35*, the barrel is exposed while all of the other pieces of the blaster are hidden.

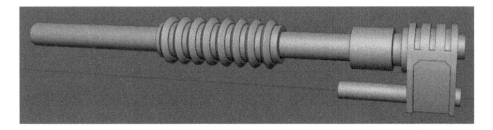

Figure 7.35 – The exposed barrel

This section is going to be our most difficult yet, as it has some small details with a lot of interacting shapes. It is important not to get overwhelmed and to just break it down into separate detailed areas that we can approach one at a time. In this case, the top of the front of the blaster in *Figure 7.36* appears to be the most detailed area, so we are going to start there.

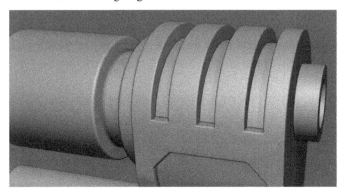

Figure 7.36 – The most detailed area on the barrel

We will approach this section, just like we have done in all of the previous sections, by outlining the edges and creases. *Figure 7.37* has all of these guiding edges in place.

Figure 7.37 – Guiding edges on the top of the blaster front

After filling all of the faces of these edges, it should look something like *Figure 7.38*.

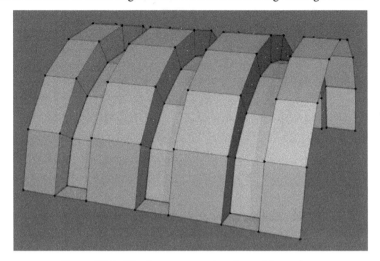

Figure 7.38 – The faces connecting the guiding edges

Now, we can move a little bit lower down to work on the indented shape in *Figure 7.39*.

Figure 7.39 – Lower detailed area

Again, we are going to begin by adding edges in the creases and edges. After outlining this shape, it should look like *Figure 7.40*.

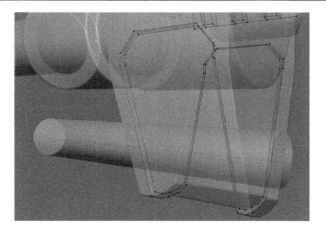

Figure 7.40 – Guiding edges on the lower part of the blaster

After we make our faces connect to everything, it should look like *Figure 7.41*.

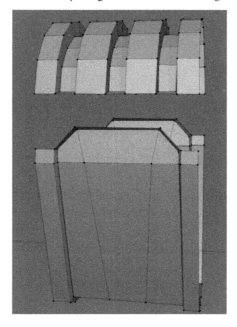

Figure 7.41 – Guiding edges filled with faces

Now, we just need to add loops to the bottom part so that it matches the top and we can connect the top and bottom of the barrel. *Figure 7.42* shows the two sections joined together.

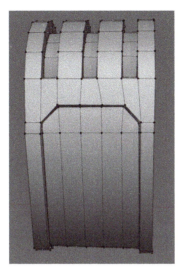

Figure 7.42 – Top and bottom joined

Now, we can add cylinders around the pipes that intersect the shape we just modeled. You can add a cylinder with *Ctrl + A*, or you can manually snap vertices around them. *Figure 7.43* shows the added cylinders.

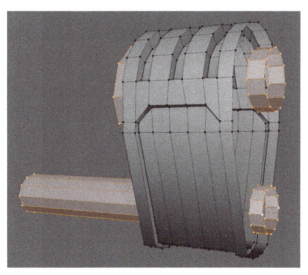

Figure 7.43 – The added barrel cylinders

Notice that the cylinders are not attached yet and this attachment is, unfortunately, quite difficult to achieve. Because the top is cylindrical and the barrel is also cylindrical, this is a good place to start. *Figure 7.44* shows these two sections connected after dissolving some of the edges to get them to match up.

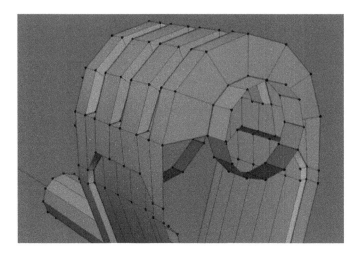

Figure 7.44 – Connecting the cylinder to the top

Next, we can join the top to the bottom like in *Figure 7.45*.

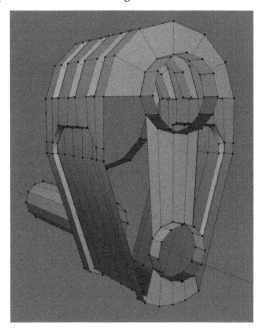

Figure 7.45 – Connecting the cylinders to each other and the bottom

We are trying to reduce the number of exposed faces on our cylinders with all of the edges we have available so that we can add as few loop cuts as possible to what we have already made. In *Figure 7.46*, we added a loop to the faces, bridging the two cylinders together, and then connected those faces to the main piece.

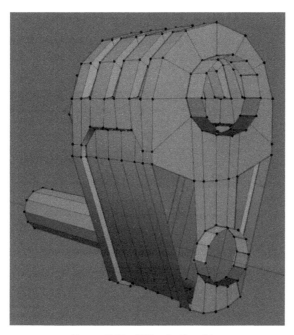

Figure 7.46 – Filling in the sides

This leaves us only needing two loops on the long faces to connect them to the bottom cylinder. The final connection is in *Figure 7.47*.

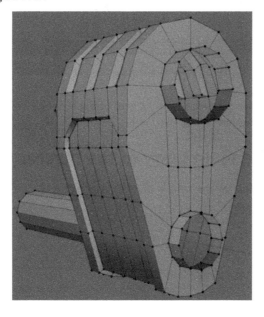

Figure 7.47 – The filled faces with the added loop cuts

After we repeat this on the backside, all that is left to finish is the barrel. Thankfully this is not too difficult to do. All we have to do is apply the mirror modifier, then extrude and scale the barrel for each step. This will be a little painful, especially at the center of the barrel with all of the ridges, but when it is done it should look like *Figure 7.48*.

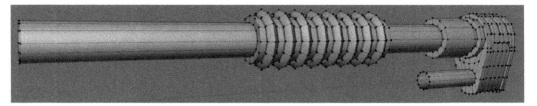

Figure 7.48 – Finished barrel

Now, we can unhide the rest of our mesh and see how well **Auto Smooth** worked, as shown in *Figure 7.49*.

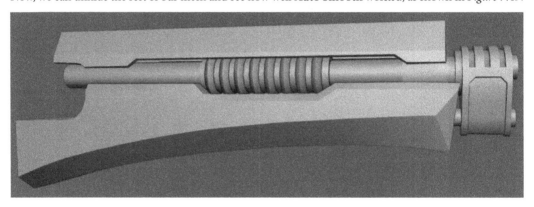

Figure 7.49 – The finished front portion of the blaster

It looks like **Auto Smooth** was able to take care of everything here because all of the edges were so sharp. So, now we can move on to the final part of the blaster, the main body.

Retopology of the main body of the blaster

Finally, we can start working on the main body of the blaster. You can get a closer look at it, with all of the other bits hidden, in *Figure 7.50*.

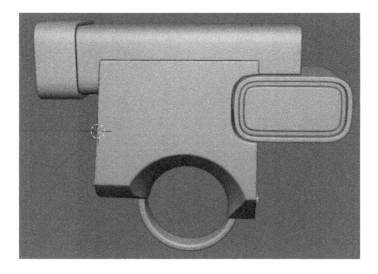

Figure 7.50 – The main body of the blaster

Figure 7.51 has all of our guiding lines drawn out for the large middle piece.

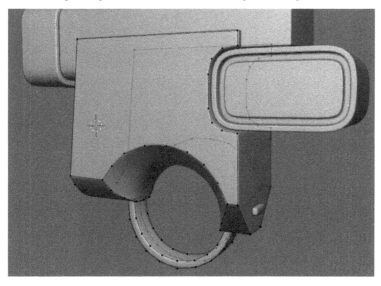

Figure 7.51 – The guiding lines on the large middle piece

Figure 7.52 shows the edges filled in with faces.

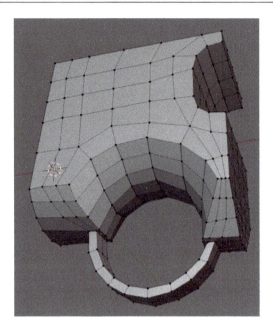

Figure 7.52 – The large middle piece filled in with faces

Now, we just repeat the same steps for the last two parts of the blaster and we are finished. *Figure 7.53* shows the final middle piece.

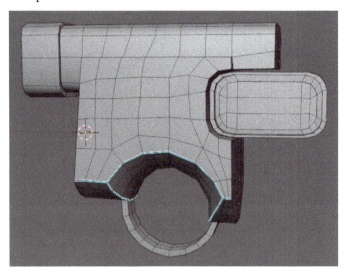

Figure 7.53 – The finished body

With that, our blaster is complete. *Figure 5.54* shows the finished blaster.

Figure 7.54 – The finished blaster

Summary

In this chapter, we learned how smooth normals work and how to use **Auto Smooth**. You should now also know how to mark edges as sharp and how to retopologize a hard surface object.

Finally, with this chapter done, we should be comfortable using the topology rules that we introduced in the first part of the book. Up until this point, we have exclusively used a quad-based workflow, only using faces that have four vertices and edges. In the next chapter, we are going to explore the use of triangles to reduce the total number of faces on a model.

8

Optimizing Geometry for a Reduced Triangle Count

All of the chapters leading up to this final chapter have been reinforcing the same ideas. Each chapter has been laying out the steps to good quad-based topology. The reason we chose this workflow was to ensure that we had a model that would be easy to work with throughout the model creation process. From deformations to UV unwrapping, quad-based topology will give you the cleanest and easiest result when done right. In this chapter, we are going to be breaking those rules.

This chapter is going to be all about reducing the triangle count on the models that we made in the last three chapters. We will start by learning why we might want to break our topology rules in the first place. Next, we will talk about some of the tools we will use to help us optimize. Then, we will go straight into optimizing a hard-surface model and finish with a character model.

In this chapter, we will be learning about the following subjects:

- Why we optimize topology
- Optimizing hard-surface meshes
- Optimizing deforming meshes

Why we optimize topology

When done correctly, quad-based topology should leave you with a model that is ready for just about everything, but having that freedom and flexibility comes with a cost. This cost is an increased number of faces on a mesh and is most evident on hard-surface models. The workflow we have been using up to this point has had us starting on the detailed areas first, then connecting those areas afterward. You may have noticed that these detailed areas have a lot of curves and complex shapes, but the areas in between that we connect them with are usually much flatter. You can see this transition in *Figure 8.1*.

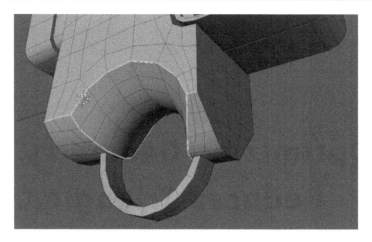

Figure 8.1 – Flatness evident in connecting areas

It is in these flat areas that there are a lot of faces that are not contributing to the shape of the blaster. That is our objective when optimizing topology; we are trying to remove as many of these unnecessary faces as possible. The question still remains: why do we want to reduce the number of faces in the first place? The answer is really quite simple. It is to reduce the performance impact of the model. This impact has two causes. The first is that the more faces there are, the more work the rendering engine is going to have to do calculating the light on each face. The second cause stems from the number of faces that the 3D software has to keep track of simultaneously. Adding a few thousand faces to a scene makes little difference on modern systems with modern software. However, when you start adding faces to multiple objects in a scene that has thousands of assets, they can start to add up quickly. That is why we are going to start by optimizing the blaster we made in the last chapter.

Optimizing hard-surface meshes

There are a few helpful tools we are going to introduce before getting into the optimization. The first tool is called **vertex slide**. Vertex slide allows us to slide a vertex along an edge. To use vertex slide, follow these steps:

1. Double-tap *G*, and you should get a line going across your edge, like in *Figure 8.2*.

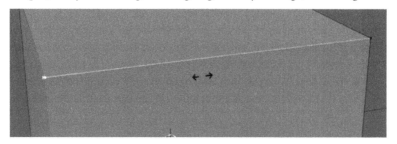

Figure 8.2 – Vertex slide initiated

2. You can now move this vertex along any of the edges that the vertex is connected to, as shown in *Figure 8.3*.

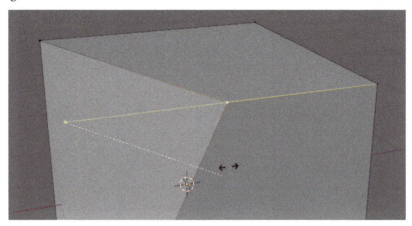

Figure 8.3 – The vertex sliding along the edge

This will blend nicely into our second tool called **Auto Merge**. **Auto Merge** merges vertices that are within a certain distance of each other. You can enable **Auto Merge** in the **Active Tool and Workspace** settings, as shown in *Figure 8.4*.

Figure 8.4 – Enabling Auto Merge

This will allow us to use vertex slide to merge vertices together at the end of an edge, like in *Figure 8.5*.

Figure 8.5 – Merging vertices at the end of the edge using Auto Merge and vertex slide

These will be the main tools we will be using to optimize our topology. We can now start to use them on our blaster. *Figure 8.6* shows us the main body, which will be where we are going to start.

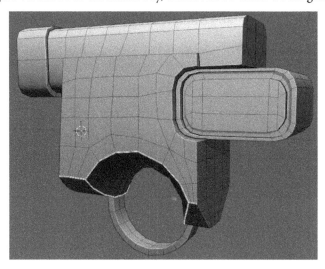

Figure 8.6 – The main body of the blaster where we will start topology optimization

The first thing we want to do is select an edge loop that we want to dissolve. This edge loop should be one that does not contribute to any of the detail of the model, like the one in *Figure 8.7*.

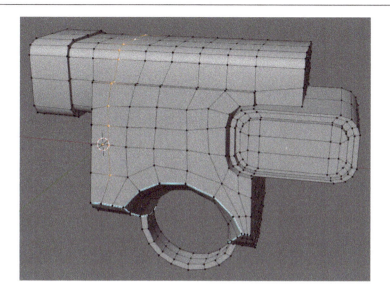

Figure 8.7 – The selected edge loop

Next, we press *G* twice to start our vertex slide, which should look like *Figure 8.8*.

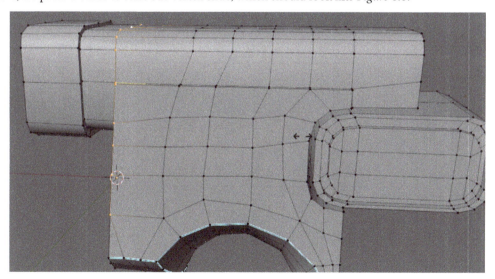

Figure 8.8 – Using vertex slide on the selected loop

Then, we will repeat this with the next row, which you can see in *Figure 8.9*.

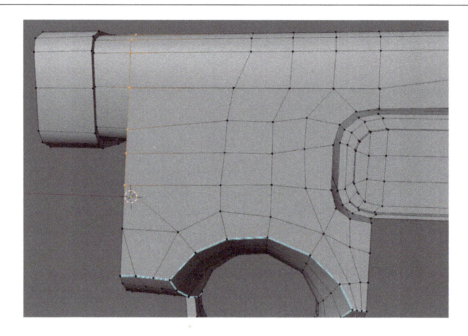

Figure 8.9 – Using vertex slide on the next row

Then, we are going to try and reduce the number of edges near the trigger guard. To do this, we just have to pull down the vertices selected in *Figure 8.10*.

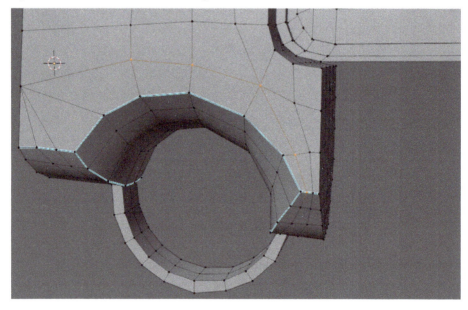

Figure 8.10 – Vertices selected on the trigger guard to be pulled down

This should result in *Figure 8.11*.

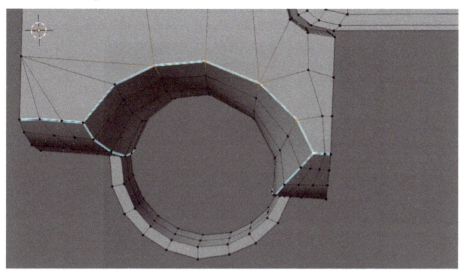

Figure 8.11 – The trigger guard after vertices have been pulled down

We want to repeat this process until all of the edges in the center of the main flap plane have been pulled to the sides. *Figure 8.12* has all of the edges on this plane pulled to the sides and merged with Auto Merge.

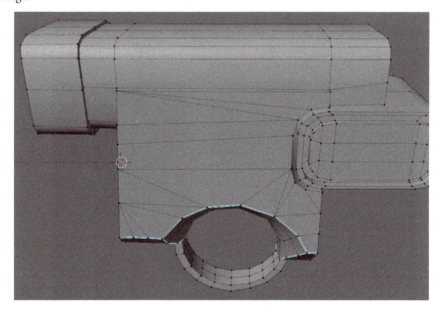

Figure 8.12 – The main plane with all edges pulled to the sides and merged

Even though we are no longer just using quads, you still want to avoid squashing the triangles as much as possible. You may have noticed that there are still quite a few edges along the sides of the plane we just optimized. That is because there are still edges on the other faces of the model, as shown in *Figure 8.13*.

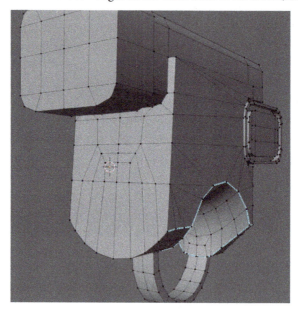

Figure 8.13 – Edges remaining on the other faces of the model

It is a good practice to deal with each side without affecting the other sides as much as possible. Then, when all of the sides are done, you can adjust those overlapping edges. In this case, we will start by pulling the vertices selected in *Figure 8.14* to the edges.

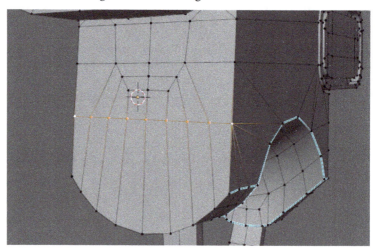

Figure 8.14 – Vertices selected to be pulled to the sides

Figure 8.15 has the vertices pulled to the side.

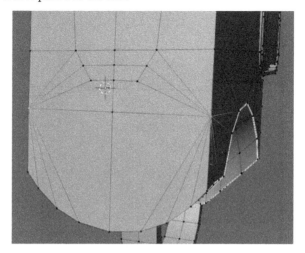

Figure 8.15 – The plane with vertices pulled to the side

Then, we will pull all of those middle vertices up into the selection in *Figure 8.16*.

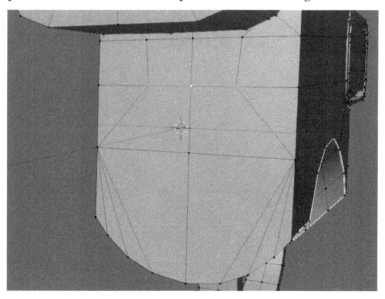

Figure 8.16 – The plane with middle vertices pulled into the selection

Then, all that is left to do is to pull the middle vertices into either the top or the bottom. In this case, it was just whichever was closer. *Figure 8.17* shows the final merging of the center vertices.

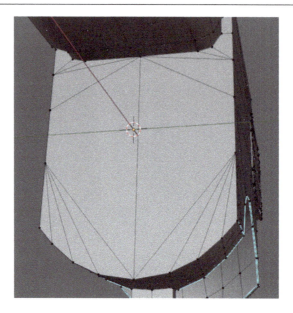

Figure 8.17 – The plane with middle vertices merged

Now, we can start to deal with the vertices on the edge. For this, we want to make sure we have a good understanding of what each of these vertices is controlling. You can see the vertices in question selected in *Figure 8.18*.

Figure 8.18 – The selected edge vertices we will work on next

Ideally, we would be able to merge all of these into one of the corners of the faces. Unfortunately, in this instance, it does not appear as though that is a possibility. If we were to merge the vertex at the bottom of this selection into the bottom corner, there would be some overlapping, as shown in *Figure 8.19*.

Figure 8.19 – Overlapping edges in the bottom corner

That is something we want to avoid, so in this case, it would be wise to keep one vertex on the edge that we can merge all other vertices to, like in *Figure 8.20*.

Figure 8.20 – A vertex retained on the edge to avoid overlapping

Now, we can take a look at the front and merge the vertices selected in *Figure 8.21* to the edges.

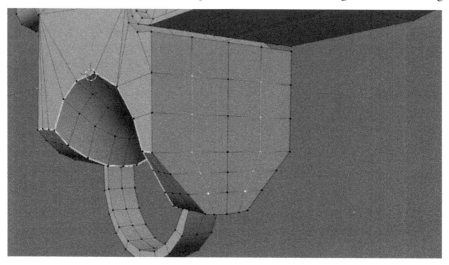

Figure 8.21 – Vertices to merge on the front

Next, we just need to extrude the selected vertices in *Figure 8.22* to the top and the bottom.

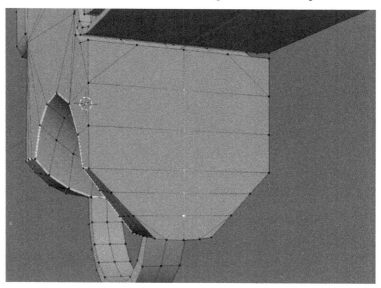

Figure 8.22 – Vertices to be merged to the top and bottom

All that is left to do is to merge the edge vertices in *Figure 8.23* to the top or the bottom.

Figure 8.23 – Edge vertices to be merged

In this case, merging them to the bottom kept any edges from getting too squashed or intersecting with each other in *Figure 8.24*.

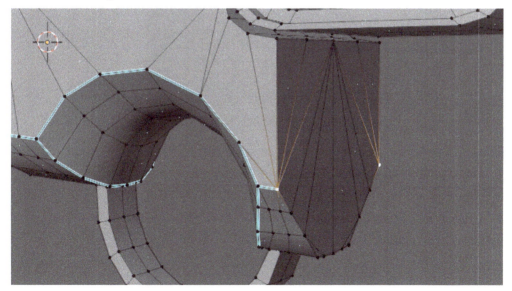

Figure 8.24 – Edge vertices after being merged to the bottom

Now, when we look at the selected vertices on the top of the model in *Figure 8.25*, we can start to merge them together.

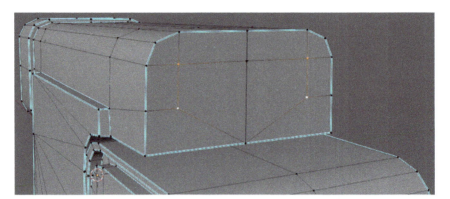

Figure 8.25 – Vertices selected at the top of the model

Figure 8.26 shows the vertices merged into the center.

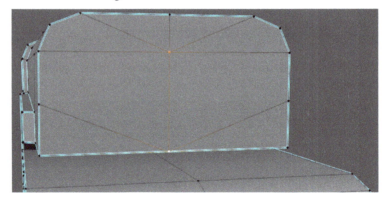

Figure 8.26 – Vertices after being merged to the center

Then, they are merged into the bottom vertex, as shown in *Figure 8.27*.

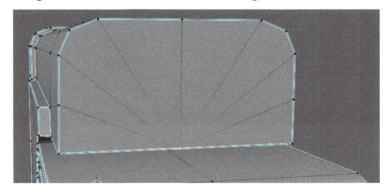

Figure 8.27 – Top vertices merged into the bottom vertex

Then, on the top of the blaster, we will merge the selected edges, as shown in *Figure 8.28*, into the edges.

Figure 8.28 – Edges to be merged on the top of the blaster

After merging, you should get a result like *Figure 8.29*.

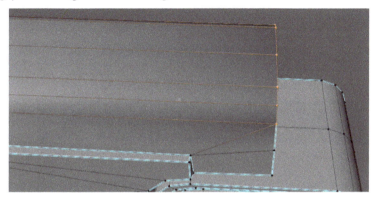

Figure 8.29 – The top of the blaster after edges have been merged

The shading has become messed up because when we merged the selected edges into the edges that were marked sharp, the sharpness was removed. This is easily remedied by marking the edges again with *Ctrl + E*, like in *Figure 8.30*.

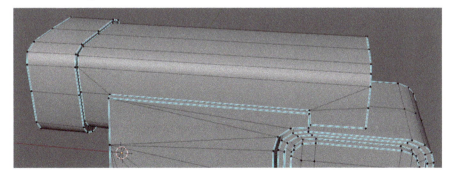

Figure 8.30 – The top blaster with edges remarked

Then, moving to the back, we have the vertices selected in *Figure 8.31*.

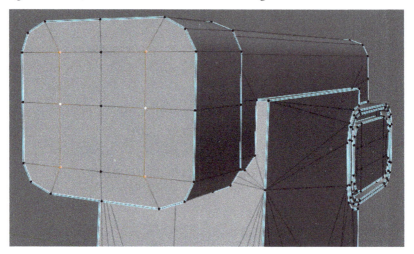

Figure 8.31 – Vertices selected at the back

Again, we are going to merge these into the center, like in *Figure 8.32*.

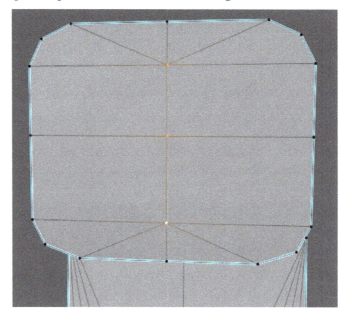

Figure 8.32 – Vertices at the back that have been merged in the center

Then, merge the remaining vertices into the top and bottom in *Figure 8.33*.

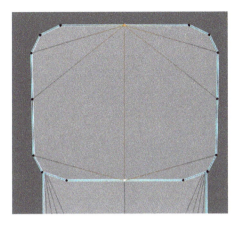

Figure 8.33 – Remaining vertices merged at the top and bottom

Now, we are going to clean up those vertices on the corners that affect multiple parts of the model, selected in *Figure 8.34*.

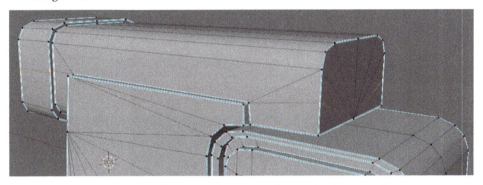

Figure 8.34 – Vertices at the corners that need to be merged

Then, as shown in *Figure 8.35*, we merge the vertices.

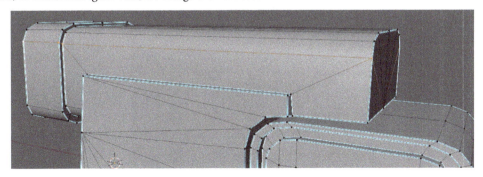

Figure 8.35 – Vertices merged at the corners

Now, all that we have left to do is our trigger guard area. To start, we are going to merge this edge into the edge above it, as shown in *Figure 8.36*.

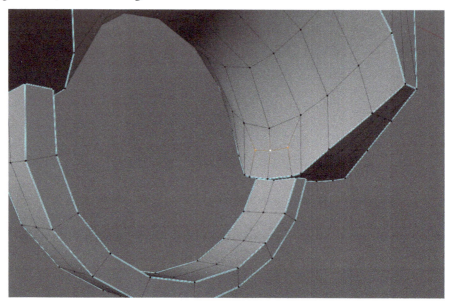

Figure 8.36 – The selected edge that needs to be merged up

You can see the result in *Figure 8.37*.

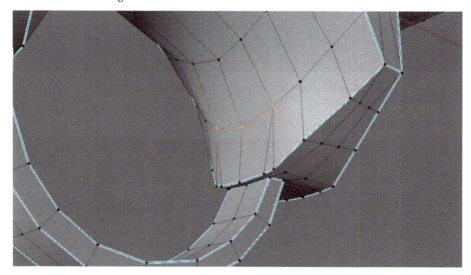

Figure 8.37 – The selected edge after merging

Now, we will look at the vertices selected in *Figure 8.38*.

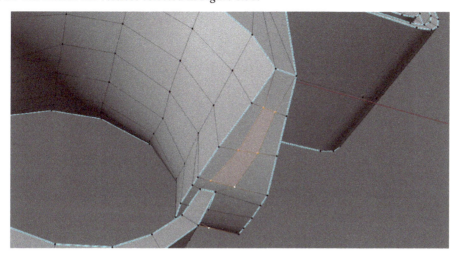

Figure 8.38 – Selected vertices

We then merge them to the left edge, like in *Figure 8.39*.

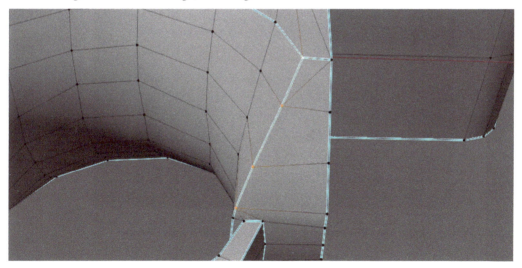

Figure 8.39 – Selected vertices merged to the left corner

To finish this off, we will merge the selected vertex on the corner between the trigger guard and the main body, as shown in *Figure 8.40*, into the vertex to the right of it.

Figure 8.40 – The vertex to be merged to the right

This is the only vertex on the edge we can merge without affecting the shape of the model. *Figure 8.41* shows the vertex after it is merged.

Figure 8.41 – The vertex on the trigger guard after being merged

The left side of the trigger is optimized the same way as this side, so when you are done, the final model should look like *Figure 8.42*.

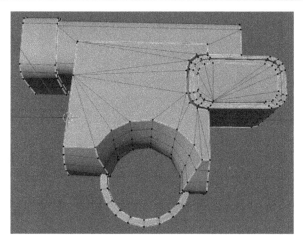

Figure 8.42 – The final model of the blaster after the trigger guard is optimized

And with that, our optimization is almost completely done. All of the other sections of the blaster were either curved or had very little to optimize because they were already very flat. *Figure 8.43* shows the final model.

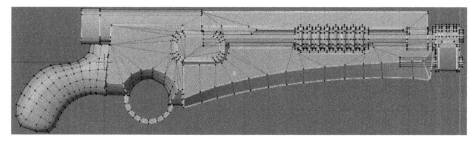

Figure 8.43 – The final complete model after optimization

Now, we can compare it to the model before we optimized it. To check the triangle count of the model, click the right mouse button while hovering over the footer at the bottom of your screen. The **Status Bar** tab should pop up. Select the **Scene Statistics** option, as shown in *Figure 8.44*.

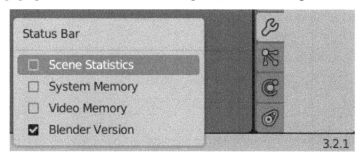

Figure 8.44 – The menu to view Scene Statistics

With that, you should be able to see the triangle count, labeled **Tris**. In this case, our model has 3,290 triangles. Our original mesh had 4,196 triangles. Therefore, almost 1,000 triangles have been removed on just this one model. *Figure 8.45* shows the two models together.

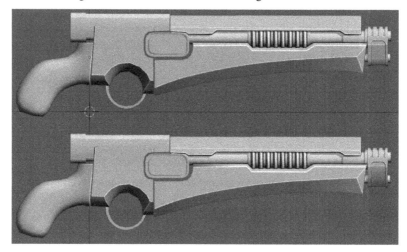

Figure 8.45 – A comparison between the unoptimized model (top) and optimized model (bottom)

The top model has substantially more triangles than the bottom one, but there is almost no visual difference between the two. The only major difference is the better performance of the optimized blaster. Now that we know the general workflow on a hard surface, we can try to adapt that to the deforming model that we made previously.

Optimizing deforming meshes

Optimizing a deforming mesh is very similar to working on a hard surface. Like before, the objective is to minimize the triangle count to improve performance. The main difference is that, now, we not only need to make sure not to disturb the shape too much but also ensure we do not affect the way the mesh deforms. That is why our strategy will be very similar. We will modify the areas of detail as little as possible, and also avoid deforming areas as much as possible. So, it is easier to start in areas that we know are not going to deform a whole lot. In this case, we will start with the side of the head shown in *Figure 8.46*.

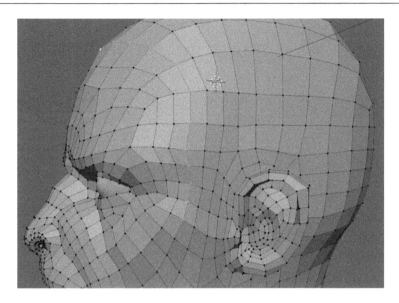

Figure 8.46 – A close-up of the side of the head that we will start work on

Like our hard-surface model, we are going to use a vertex slide to merge these vertices into the row next to them. You can see the merged vertices in *Figure 8.47*.

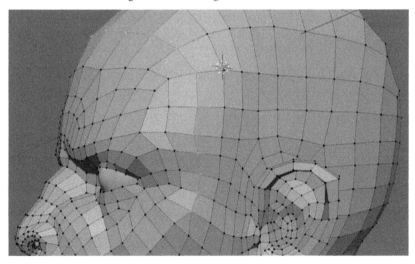

Figure 8.47 – The first row of merged vertices on the side of the head

Then, we will repeat this for the next row to the right of it, which you can see in *Figure 8.48*.

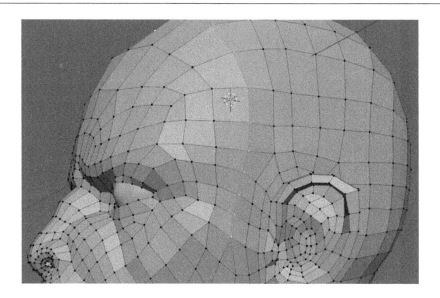

Figure 8.48 – The second row of merged vertices on the side of the head

Note how there are now two triangles at the base of the merges that we just did. If we take a closer look at *Figure 8.49*, we can see that this merge also created a pole connecting five edges.

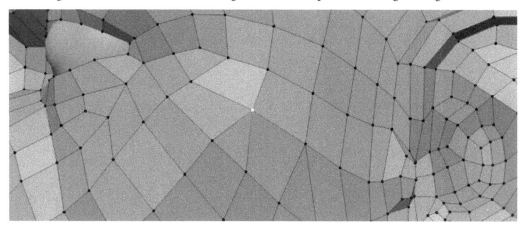

Figure 8.49 – A pole connecting five edges

We want to make sure that when we merge things together, we do not form poles with more than five edges. We are going to continue merging edge loops along the back of the head, starting above the ear, as shown in *Figure 8.50*.

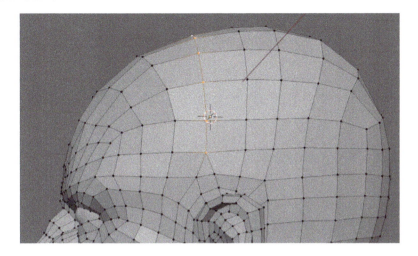

Figure 8.50 – The area above the ear merged

Then, we are going to merge two more at the back of the head in *Figure 8.51*.

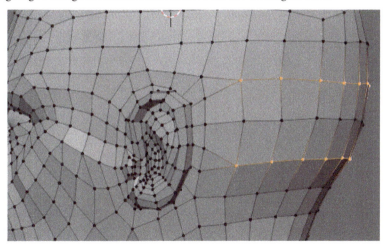

Figure 8.51 – The area at the back of the head merged

Then, we want to look at loops going in the other direction across the head, such as the loop merged in *Figure 8.52*.

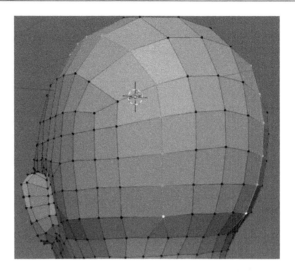

Figure 8.52 – The merged loop at the back of the head

This same merge also wraps around the front face, as shown in *Figure 8.53*.

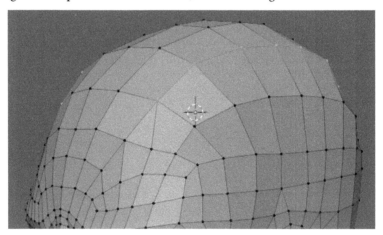

Figure 8.53 – The merged loop from the back of the head also merged at the front

Note how the triangle starts above where most of the animating will be done for the brow. Now, we can start working on the neck. *Figure 8.54* shows the merging we are going to be doing across the neck.

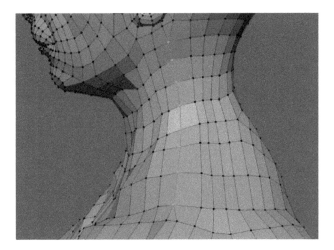

Figure 8.54 – The area on the neck merged

Figure 8.55 shows the topology after merging.

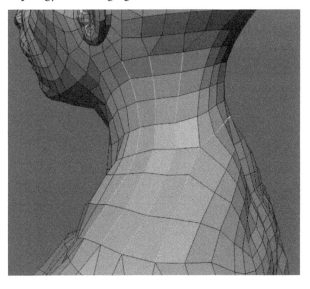

Figure 8.55 – The topology of the neck after merging

All of the triangles are far enough away from the neck to not affect the deformations at all. On the front of the neck, we merged two edge loops together, going all the way from the chin to the bottom of the chest, as shown in *Figure 8.56*.

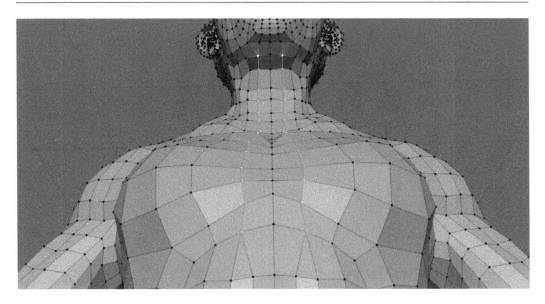

Figure 8.56 – Front of the neck after being merged

With that, we can move on to the back. *Figure 8.57* shows us the neck, which now looks much better.

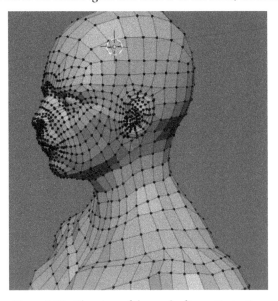

Figure 8.57 – The view of the neck after optimization

The back has one major area at the top around the selected edges, as shown in *Figure 8.58*, that we need to address.

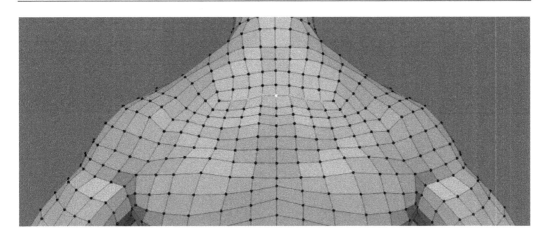

Figure 8.58 – The area on the back that needs to be optimized

We merge this area into three edges, as shown in *Figure 8.59*.

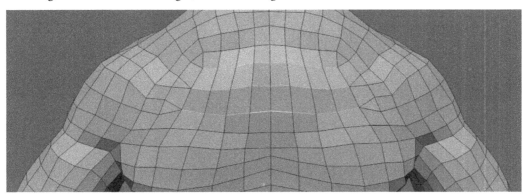

Figure 8.59 – The area on the back after the edges have been merged

After dealing with this, the back looks pretty good, so next, we will move on to the arm. On the arm, we do not want to reduce the number of edge loops going around the arm, as these are most important for twisting and bending deformations on a cylinder-like object such as our arm. We are going to merge edges like those shown in *Figure 8.60*.

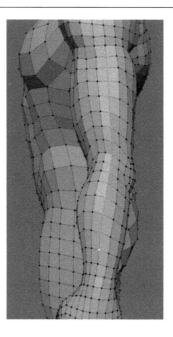

Figure 8.60 – Edges to be merged on the arm

This way, we can put the triangles our merging makes before the wrist and the shoulder. It ends up that every other pair of edge loops can be merged, like *Figure 8.61*.

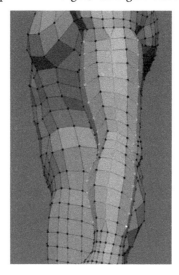

Figure 8.61 – The arm after being merged

Now, we just have to repeat this process on the legs, like in *Figure 8.62*.

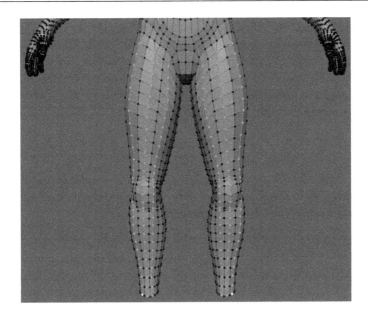

Figure 8.62 – Legs after optimization

And with that, our optimization is done. *Figure 8.63* shows the optimized mesh next to the original mesh.

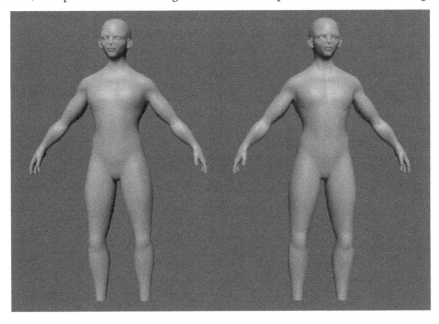

Figure 8.63 – Comparison of the optimized mesh (left) and the original mesh (right)

Again, I think it would be hard to tell which model was optimized and which one was not. The model on the left is the optimized model with 10,240 triangles, and the model on the right is the original mesh at 11,976. That is a change of over 1,700 triangles with just merging a few edge loops. There is definitely room to reduce the triangle count by a few more, but there is a point where you are no longer removing enough triangles to justify the amount of time spent on optimizing. With this model, we have reached a satisfactory point, with relatively little effort.

Summary

In this chapter, we learned why optimizing topology is important. We also explored how to achieve this optimization by using Auto Merge and vertex slide. By now, you should have a good idea of how to optimize hard-surface models and deforming models.

With this, our topology process is complete. Now that we have our optimized meshes, they are ready for rigging, materials, and then finally, whatever medium they are getting used for. From the fundamentals of topology to the retopology of complex hard-surface and deforming meshes, you should be well-equipped to approach them all as you develop your own workflow based on this foundation.

Index

V

W

Other Books You May Enjoy

If you enjoyed this book, you may be interested in these other books by Packt:

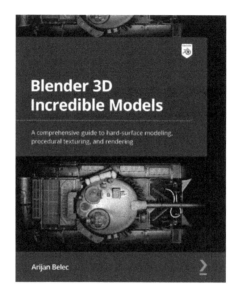

Blender 3D Incredible Models

Arijan Belec

ISBN: 9781801817813

- Dive into the fundamental theory behind hard-surface modeling
- Explore Blender's extensive modeling tools and features
- Use references to produce sophisticated and accurate models
- Create models with realistic textures and materials

Shading, Lighting, and Rendering with Blender EEVEE

Samantha Crowder

ISBN: 9781803230962

- Explore EEVEE Render Properties for optimal outcomes
- Focus on shading processes, including those that are both traditional and more cutting-edge
- Understand composition and create effective concept art inside Blender
- Discover procedural workflows to shorten the artistic process instead of getting mired in details

Packt is searching for authors like you

If you're interested in becoming an author for Packt, please visit `authors.packtpub.com` and apply today. We have worked with thousands of developers and tech professionals, just like you, to help them share their insight with the global tech community. You can make a general application, apply for a specific hot topic that we are recruiting an author for, or submit your own idea.

Share Your Thoughts

Now you've finished *Squeaky Clean Topology in Blender*, we'd love to hear your thoughts! Scan the QR code below to go straight to the Amazon review page for this book and share your feedback or leave a review on the site that you purchased it from.

`https://packt.link/r/1-803-24408-9`

Your review is important to us and the tech community and will help us make sure we're delivering excellent quality content.

Download a free PDF copy of this book

Thanks for purchasing this book!

Do you like to read on the go but are unable to carry your print books everywhere?

Is your eBook purchase not compatible with the device of your choice?

Don't worry, now with every Packt book you get a DRM-free PDF version of that book at no cost.

Read anywhere, any place, on any device. Search, copy, and paste code from your favorite technical books directly into your application.

The perks don't stop there, you can get exclusive access to discounts, newsletters, and great free content in your inbox daily

Follow these simple steps to get the benefits:

1. Scan the QR code or visit the link below

https://packt.link/free-ebook/9781803238975

2. Submit your proof of purchase
3. That's it! We'll send your free PDF and other benefits to your email directly

www.ingramcontent.com/pod-product-compliance
Lightning Source LLC
Chambersburg PA
CBHW060542060326
40690CB00017B/3575